SpringerBriefs in Energy

More information about this series at http://www.springer.com/series/8903

Zainal Ambri Abdul Karim
Shaharin Anwar Bin Sulaiman
Editors

Alternative Fuels for Compression Ignition Engines

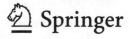 Springer

Editors
Zainal Ambri Abdul Karim
Department of Mechanical Engineering
Universiti Teknologi PETRONAS
Seri Iskandar, Perak
Malaysia

Shaharin Anwar Bin Sulaiman
Department of Mechanical Engineering
Universiti Teknologi PETRONAS
Seri Iskandar, Perak
Malaysia

ISSN 2191-5520   ISSN 2191-5539 (electronic)
SpringerBriefs in Energy
ISBN 978-981-10-7753-1   ISBN 978-981-10-7754-8   (eBook)
https://doi.org/10.1007/978-981-10-7754-8

Library of Congress Control Number: 2017964254

Printed on acid-free paper

This Springer imprint is published by the registered company Springer Nature Singapore Pte Ltd. part of Springer Nature
The registered company address is: 152 Beach Road, #21-01/04 Gateway East, Singapore 189721, Singapore

Preface

The chapters contributed in this book are the collections of several groups of research works undertaken at the Centre for Automotive Research and Electric Mobility (CAREM), Universiti Teknologi PETRONAS, Malaysia. This is an initial effort to amass all knowledge gained and experiential learning from the postgraduate students, researchers, and academic staff in developing the experimental methods, rigs, and test beds in order to accomplish the research endeavor into a book series.

This current book is entitled *Alternative Fuels for Compression Ignition Engines*, aptly to address one of the major research focuses of CAREM, which is to enhance the development and utilization of alternative fuel, in order to reduce or control the environmental impact of internal combustion engine exhaust gases. The alternative fuels discussed in this book include dual fuel techniques, rubber seed/palm oil biodiesel, syngas dual-fueling, water-in-diesel emulsion, and gasification of date palm seeds. The editors hope that this book on alternative fuels for diesel engine will be of benefit to researchers in the field of engine development and on alternative fuels.

The Centre for Automotive Research and Electric Mobility (CAREM) was conceptualized from the outcome of Universiti Teknologi PETRONAS (UTP) research and development Master Plan in 2002 which outlines transportation as one of the focus areas of R&D for UTP. The center was set up on August 15, 2005, from the fund of two major grants under the IRPA-PR from Ministry of Science, Technology and the Innovation (MOSTI) in 2003. The vision of CAREM is to emerge as a leader in energy management in transportation and novel power generation, while the mission of CAREM is (1) to provide opportunities for the pursuit of knowledge and expertise in the advancement of energy management in transportation and novel power generation, (2) to perform world-class research and produce technological products with the potential to compete in the world market, and (3) to nurture creativity and innovativeness among team members by providing support in terms of expertise, facilities, and funds. The end state is to create a world-class center of excellence (COE) for novel energy management in transportation and power generation.

Seri Iskandar, Malaysia   Zainal Ambri Abdul Karim
                                                            Shaharin Anwar Bin Sulaiman

Acknowledgements

The editors wish to acknowledge with great appreciation to the following authors that have contributed chapters to this book publication.

Rasheed Adewale Opatola
A. Rashid A. Aziz
Morgan R. Heikal
Mior Azman Meor Said
Ibrahim Khalil Adam
Suzana Yusup
Bahaaddein K. M. Mahgoub
Shaharin A. Sulaiman
Z. A. Abdul Karim
Mohammed Yahaya Khan
Mohammed Elamen Babiker

The research works presented in this book have been made possible by the following research grants.

MyRA—0153AB-J17
URIF—11/2013
URIF—6/2012

We are also grateful to the services and support provided by the laboratories for the usage of the facilities and the personnel whose expertise is invaluable.

Centre for Automotive Research and Electric Mobility (CAREM)
Dr. Firmansyah, Manager, Advanced Propulsion
Mr. Ezrann Zharif Zainal Abidin, Manager, Smart Mobility
Ahmad Shahrul
Mahfuzrazi

Centre for Biofuel and Biochemical Research (CBBR)

Contents

About the Editors

Dr. Zainal Ambri Abdul Karim currently serves as an Associate Professor at the Mechanical Engineering Department of Universiti Teknologi PETRONAS, Malaysia, and has been with the department for over 20 years. He obtained his B.Sc. in Marine Engineering (California Maritime Academy, USA), earned his M.Sc. in Marine Engineering from University of Newcastle Upon Tyne (UK), and later read Automotive Engineering to Ph.D. level at Loughborough University (UK). He had conducted several short courses for the Accelerated Capability Development for PETRONAS Engineers Skill Group-2 which include Design, Operation, Maintenance and Inspection of Steam Boilers, Internal Combustion Engines, and Automotive Engineering. In addition, he also contributed to other short courses such as Centralized Distributed Cooling System and Thermal Power Plant Efficiency and Heat Rate Improvement. He is an active reviewer for several international journals while having published more than 70 journal articles. His research interest includes combustion analysis, energy optimization, marine engineering, power plant engineering, and internal combustion engines.

Dr. Shaharin Anwar Bin Sulaiman is an Associate Professor in the Department of Mechanical Engineering at Universiti Teknologi PETRONAS, Malaysia. He graduated in 1993 with B.Sc. in Mechanical Engineering from Iowa State University. He earned his M.Sc. in Thermal Power and Fluids Engineering from UMIST in 2000 and obtained his Ph.D. in Aerosol Combustion from the University of Leeds in 2006. During his early years as a graduate, he worked as a Mechanical and Electrical (M&E) Project Engineer. His research interests include air-conditioning and ventilation, combustion, sprays and atomization, and biomass energy. He has published more than 130 journal articles in various engineering researches. He is a Member of the ASHRAE and a Corporate Member of the Institution of Engineers Malaysia (IEM). Presently, he is Director of the Hybrid Energy Systems research group in the university and also certified Professional Engineer under the Board of Engineers Malaysia.

Chapter 1
Dual Fuel (Gas–Liquid Diesel)

Rasheed Adewale Opatola, A. Rashid A. Aziz, Morgan R. Heikal
and Mior Azman Meor Said

1.1 Introduction

Diesel engines constitute an integral part of global transportation and industrial infrastructures, particularly in heavy-duty applications such as trucks, buses, farm equipment, locomotives, and ships due to their high thermal efficiency and durability. Exhaust emissions from diesel engines are part of environmental pollutants, particularly oxides of nitrogen (NO_x), particulate matter (PM), unburnt hydrocarbon (HC) and carbon monoxide (CO), with potential health concerns [1]. The positive achievements in terms of efficiency associated with diesel engines are, therefore, overshadowed by its high emissions drawback.

Although, advances in technical know-how have resulted in the advent of novel combustion concepts aimed at reducing NO_x and PM emissions, nevertheless, there are still room for further exploits into new and customised solutions to the hazards of diesel engine exhaust emissions. Hence, the need for an appropriate technology that can be applied to new and existing engines, which would enhance simultaneous reduction of NO_x and PM emissions from diesel engines. In this chapter, two of the emerging techniques for mitigating exhaust emissions of diesel engines are discussed.

R. A. Opatola (✉) · A. R. A. Aziz · M. A. Meor Said
Mechanical Engineering Department, Centre for Automotive Research and Electric Mobility,
Universiti Teknologi PETRONAS, 32610 Seri Iskandar, Perak, Malaysia
e-mail: surulere_opatola@hotmail.com

A. R. A. Aziz
e-mail: rashid@utp.edu.my

M. A. Meor Said
e-mail: miorazman@utp.edu.my

M. R. Heikal
School of Computing, Engineering and Mathematics, Advanced Engineering Centre, University
of Brighton, Brighton BN2 4GJ, UK
e-mail: m.r.heikal@brighton.ac.uk

© The Author(s), under exclusive licence to Springer Nature Singapore Pte Ltd.,
part of Springer Nature 2018
Z. A. Abdul Karim and S. A. Sulaiman (eds.), *Alternative Fuels for Compression
Ignition Engines*, SpringerBriefs in Energy, https://doi.org/10.1007/978-981-10-7754-8_1

The first technique, the combustion-based approach, moderates exhaust emissions through improvements to the combustion processes. The second method is the fuel-based approach, which often considers modification of the fuel itself to control the combustion processes and hence emissions.

The demise of Rudolf Diesel in 1913 saw his engine modified to run on the polluting petroleum fuel, now known as diesel [2, 3]. Today, over a century of development by some of the best engineers have made the diesel engine a highly-refined machine; and the rising costs of fossil fuel have brought it into sharp focus as the technology that will define the present century as well [4].

1.2   Diesel Engine Combustion and Emissions

The diesel engine is often supplied with fuel from the injection nozzle at a pressure of 10–35 MPa for indirect injection engines, and up to 100 MPa or higher for direct injection engines. The high pressure is vastly essential to deliver fuel against the highly compressed air in the engine cylinders at the end of the compression stroke, and to break up the fuel oil which has low volatility and is often viscous. As a fossil fuel, diesel is a mixture of complex hydrocarbons that can produce harmful gases such as NO_x and soot during combustion. The cause of these noxious emissions can be incomplete combustion of the fuel, reaction between mixture components under high temperature and pressure, combustion of engine lubricating oil and combustion of non-hydrocarbon components in the fuel [5]. In order to burn, diesel fuel must first be in a vapour state. This process, called vaporisation, is enriched if the liquid fuel is already sprinkled in small droplets. The mixture strength limits that are suitable for auto-ignition can be estimated through the study of flammability limits; for diesel oil, the range is 0.6–5.5% fuel by volume [5], corresponding to a range of fuel-air equivalence ratio of 0.3–3.5. The inevitable by-products of the combustion of fuel and air are strongly dependent upon, firstly, the charge-air, its temperature, pressure, motion; secondly, the combustible fuel, its type, injection, atomisation, evaporation; and thirdly, their states and interaction, leading to the auto-ignition and combustion of the charge [6].

Usually, the formation mechanisms of HC, CO, PM and NO_x in the combustion chamber strongly depend on temperature, local concentration of oxygen and duration of combustion as well; the HC, CO and PM increase when temperature drops, while the NO_x increases as temperature increases [5]. Unburnt hydrocarbon (HC) emissions emerge through two probable means in a diesel engine. Firstly, when the local fuel and air blend is too lean to either burst into flames or sustain a spreading flame; and secondly, when the fuel and air mixture is too rich to ignite or sustain a burning flame. This accounts for the HC emissions witnessed with recent advances in diesel combustion. The formation of CO depends on the availability of unburnt gaseous fuel and mixture temperature, which regulate the rate of fuel decomposition and oxidation. Hence, CO emissions are produced in combustion regions where the air-fuel ratio (AFR) is close to stoichiometric or below, reflecting a lack of oxygen [5].

This has hardly been an issue with conventional diesel engines because even in the most awful instances, the flame is stabilised at a mixture where air-fuel ratio is above stoichiometric.

PM emerge through the pyrolysis of fuel by surrounding hot gases within the flame region at temperatures between 1000 and 2800 K. The diffusion-combustion phase is the main contributor to PM emissions. Usually, premixed combustion is characterised by high levels of mixing, making it too lean for PM formation. Nonetheless, if the temperature of diffusion-combustion phase is high enough, it will promote PM oxidation and reduce the total emissions. NO_x emissions are produced at high temperatures during combustion. These conditions are attained during the combustion of a close-to-stoichiometric mixture in a high pressure and temperature environment. A direct correlation has been established between the formation rates of NO_x and the flame temperatures in combustion above 700 K. Premixed combustion will not contribute much to NO_x emissions because the mixture is well mixed at that stage and the temperatures are quite low. The extended Zeldovich mechanism, described by Heywood [5], is generally used in computational fluid dynamics to model NO_x emissions.

1.3   European Union Emission Standards

Emission standards are the instruments utilised in controlling ambient pollutants. These standards fix numerical restrictions on the allowable quantity of air pollutants that may be released from specified sources over explicit durations. They are generally intended to improve air quality standards and protect human health. While, several emissions criteria regulate noxious waste emitted by automobiles (motor cars) and other power-driven vehicles; others set limits on emissions from industry, power plants, small equipment such as lawn mowers and diesel generators, and other sources of air pollution. Presently, standards are in place for all road vehicles, trains, ferries and non-road mobile machinery such as tractors. However, there are no standards yet for maritime ships and aircrafts. The foremost standards, now known as Euro 1, for emissions from new passenger cars and light-duty vehicles was introduced in July 1992 as EC93 and since then the European Union (EU) has progressively tightened these standards through the institution of Euro 2, 3 and 4 [7]. In further pursuance of acceptable air quality, the EU instituted a significantly tighter emission limits, known as Euro 5 and Euro 6 emission standards to regulate atmospheric pollutants such as PM, NO_x and HC for vehicles sold in the EU market. The main effect of Euro 5 over Euro 4 was to tighten the emission of particulate matter from diesel cars from 25 to 5 mg/km, which necessitated the introduction of particle filters for diesel cars [8–10]. Euro 6 was targeted at reducing NO_x emissions from diesel cars further, from 180 to 80 mg/kg. Evidently, the main targets of these directives were NO_x and PM emissions from diesel passenger cars since their emission limits were projected to reduce by 60 and 90% respectively from Euro 3 to proposed Euro 5.

More recently, the European Commission performed a detailed analysis and found that emissions generated by real driving on the road of Euro 5/6 vehicles significantly exceed the emissions measured on the regulatory new European driving cycle (NEDC), in particular with respect to NO_x emissions of diesel vehicles [11]. Subsequently, the regulation of 2008 as regards emissions from light passenger and commercial vehicles (Euro 6) was amended in 2016 by the EU [11]. The amended regulation contains the revised procedures, tests and requirements for type-approval as set out in Regulation (EC) No 692/2008 so that they satisfactorily reflect the emissions generated by real driving on the road. Hence, a real driving emission (RDE) test procedure has now become mandatory for type-approval of motor vehicles with respect to emissions from light passenger and commercial vehicles (Euro 5 and Euro 6). The Commission chose 1 January 2016 as the effective date in order to allow stakeholders and manufacturers ample time to comply with the requirements of the new Regulation.

The regulation on carbon dioxide (CO_2) emissions, which at inception basically covers passenger cars and vans [12, 13], has been recently extended to heavy-duty vehicles (HDVs) [14]. The CO_2 directive differs from previous Euro emission standards in that compliance is not required for a single vehicle but for the weighted performance of the entire fleet produced by a manufacturer in a year. The first binding limits for CO_2 emissions from vehicles were approved in 2009, when the EU set a legally binding CO_2 standard for new cars [12]. Monitoring of the mandatory CO_2 emission targets (2015 and 2020) for new passenger cars started in 2010. There is currently no after-treatment technology that can reduce CO_2 emissions from road vehicles, hence, the standards can also be seen as fuel efficiency standards. The CO_2 emission standard for cars is designed to ensure that the average car sold in Europe should not emit more than 130 and 95 g/km of CO_2 by the years 2015 and 2020 respectively [15]. Furthermore, legislations are in force to control CO_2 emission from new vans (light commercial vehicles) sold on the European market. The law stipulates that new vans registered in the EU should not emit more than an average of 175 g/km of CO_2 by 2017 [13]. As lately amended, the target for 2020 is 147 g/km of CO_2 for the average emissions of new light commercial vehicles registered in the Union, as measured in accordance with Regulation (EC) No 715/2007 [16]. The 2020 target corresponds to about 5.5 l/100 km of diesel consumption. It is noteworthy that in 2014, in order to fine-tune the modalities for reaching the 2020 target to reduce CO_2 emissions from new passenger cars and light commercial vehicles, there were amendments to the previous regulations of 2009 and 2011 [16, 17].

Lately, the European Commission has instituted a strategy to curtail CO_2 emissions from HDVs. The HDV strategy, adopted in May 2014, is the first step by the EU to tackle CO_2 emissions from trucks, buses and coaches [14]. The Commission may consider supplementary measures to curtail CO_2 emissions from HDVs when this legislation is enforced. The most likely route may be to set compulsory limits on average CO_2 emissions from newly-registered HDVs, as is currently in force for cars and vans.

Beijing introduced the Euro IV standard in advance on January 1, 2008; thus becoming the first city in mainland China to adopt this standard [18]. South Africa's first clean fuels programme was implemented in 2006 with the banning of lead from petrol and reduction of sulphur levels in diesel from 3000 to 500 parts per million (ppm), along with a niche grade of 50 ppm. The Clean Fuels 2 standard, expected to begin in 2017, includes reduction of sulphur to 10 ppm; lowering of benzene from 5 to 1% of volume; reduction of aromatics from 50 to 35% of volume; and the pegging of olefins at 18% of volume. The prime inference is that emission regulations are here to stay, and from all indications they will only get more stringent.

1.4 Emerging Trends in Mitigating Automotive Exhaust Emissions

Expectedly, there have been upsurges in research and development in engine technology aimed at reducing emissions from diesel engines. Modern hardware-based solutions such as advanced fuel injection equipment, piezo-injectors, injector nozzle modification, after-treatment devices are being explored by automotive engineers [19]. Although, these technologies give encouraging results, they come with high price tags. Moreover, retrofitting them into existing engines is often a daunting task. Their cost and complexity, therefore, threaten the competitiveness of the diesel engine package. There are other options available to the automotive industry in complying with these emissions regulations such as particulate filters for PM emissions, NO_x traps for NO_x emissions, and selective catalytic reduction (SCR) for NO_x, HC and PM emissions. Moreover, electric, fuel-cell and solar vehicles are all possible means of getting round the emission legislations. However, constraints in terms of infrastructures, range, performance and comfort are issues that need to be addressed.

The lean burn technique, which often results in reduced in-cylinder temperature and low thermal stresses in gasoline engines, was one of the prior methods used in controlling NO_x emissions. It allows the use of turbo-charging, high compression ratio, and optimum spark advance timing to achieve high engine efficiency. Although it fulfilled previous emission standards without the need for exhaust gas after-treatment resulting in lower engine costs, stringent emission standards have rendered it almost obsolete [20]. Further cutting-edge researches have yielded the emergence of the low temperature combustion techniques, which aimed at improved fuel consumption and reduced engine-out emissions through modification of combustion processes [21]. These include homogeneous charge compression ignition (HCCI), premixed charge compression ignition (PCCI), partly-premixed compression ignition (PPCI) and exhaust gas recirculation (EGR). HCCI and PCCI engines demonstrate pronounced capabilities in terms of simultaneous ultra-low NO_x and PM emissions. However, the HCCI engines are inhibited by obstacles such as combustion control, high HC and CO emissions at low loads, increase in NO_x emission at high loads, high rates of heat release [22, 23]. PCCI, on other hand, is difficult to achieve at high engine loads; this fact combined with cold start challenges and the higher CO and HC emissions lead to a scenario where the use of catalysts becomes imperative in PCCI operation [24]. Hence, commercial applications of HCCI and allied techniques in diesel engines are presently not feasible [25]. Other methods that have been studied for the mitigation of NO_x emissions are retarding injection timing, water injection, multi-stage injection and EGR [5, 26, 27].

Recently, dual-fuel combustion technologies, which often consider modification of the fuel itself in controlling the combustion processes; and hence engine-out emissions, have elicited extensive researches and applications. Alternative fuel solutions such as compressed natural gas, liquid petroleum gas, hydrogen (H_2), emulsified, bio or wide-cut fuels have been implemented in dual-fuel engine vehicles (with diesel or gasoline). They are, however, faced with a barely developed infrastructure, storage and high-cost of fuel, and above all, do not offer the same level of benefits as the alternate powertrain solutions, as identified by Garvine [28]. Therefore, this limits their application to contained fleets such as public-transport, airport or agricultural vehicles. Nonetheless, natural gas appears more promising because of cleaner combustion, availability and simplicity in applications. For instance, studies have revealed that, in addition to improved fuel economy, the dual-fuel combustion concept utilising compressed natural gas and diesel (CNG-Diesel) is found supportive in the mitigation of pollutant emissions from diesel engines [29, 30].

1.5   Combustion-Based Emissions Control Techniques

Generally, the prevalent methods employed in mitigating engine exhaust emissions can be categorised as hardware-based, fuel-based and combustion-based techniques. In spite of the prominence of the hardware-based technique, the other two methods, discussed in this chapter, have recently elicited attention amidst researchers. The combustion-based technique, which moderates exhaust emissions through improvements to combustion processes, can be categorised as low temperature combustion and exhaust gas recirculation.

1.5.1   Low Temperature Combustion

Researchers have tried to extenuate exhaust emissions through low temperature combustion (LTC) technique [21] such as HCCI [31], PCCI [32], PPCI [33] and EGR [34]. HCCI is as a premixed, lean burn combustion process [31], that utilises long ignition delay to achieve homogeneous mixture before auto ignition, avoiding the problems of non-homogeneous mixture combustion [35]. Ghazikhani et al. [36] investigated the effect of premixed and equivalence ratios on CO and HC emissions of emissions of gasoline-diesel dual-fuel HCCI engine. Their finding indicated an increase in HC emissions by increasing equivalence and premixed ratios; they ascribed this outcome to the crevices and quenching phenomenon near the cylinder wall, which prevent oxidation of mixture during low temperature HCCI combustion. There was, however, a decrease in CO emission due to more conversion of CO to CO_2 reactions.

Kobayashi et al. [37] studied the impacts of a turbo-charged natural gas HCCI engine on combustion and emissions. The study revealed that thermal efficiency was improved by raising the engine compression ratio and lowering the boost pressure. For instance, at an engine compression ratio of 21 and turbo-charging pressure of 0.19 MPa, a brake thermal efficiency of 43% was attained; the NO_x emissions were only 10 ppm or less. They further examined the performance of the engine fitted with a newly developed turbocharger. As a result, 43.3% brake thermal efficiency, 0.98 MPa brake mean effective pressure, and 13.8 ppm NO_x emissions were realised. They reported a NO_x emission factor of 0.096 g/kWh, a tremendously low value and thus concluded that it was not necessary to install exhaust treatment equipment. In their work on the evaluation of HCCI for future gasoline powertrains, Osborne et al. [38] reported that HCCI operation yielded 99% reduction in NO_x and an 8% reduction in indicated specific fuel consumption (*isfc*) when compared with baseline direct injection gasoline engine operation mode. HC emissions for HCCI operation were found to be comparable to other DI gasoline engine modes.

However, there are several problems to be solved before the commercial application of HCCI becomes feasible. Whereas, it is crucial to control combustion for best fuel economy and ultra-low emissions; ironically, HCCI engine does not have a direct mode to control the ignition timing. Other hurdles include difficulty in extending the operating range of HCCI combustion to high loads, transient response of HCCI engine, high HC and CO emissions mostly at low loads, high rates of heat release and increase in NO_x emission at high loads [39, 40]. Although, the viability of a full fledge HCCI diesel engine had been established in a single-cylinder heavy-duty Caterpillar engine, which attained 20 bar *bmep* [41]; nevertheless, the extension of full load HCCI application to light duty engines remains unclear [25].

The partially premixed compression ignition (PPCI) method has also elicited the attention of researchers. Mixture preparation for PPCI usually occurs through direct injection of the fuel into the cylinder. The effects of injection characteristics such as injection duration, spray angle, nozzle hole diameter, injected fuel temperature and EGR rate were investigated on combustion and emissions in a gasoline fuelled PPCI engine [33]. The results indicated that optimisation of injection characteristics led to simultaneous reduction in NO_x and soot emissions with negligible change in indicated mean effective pressure (*imep*). This was because, increasing the injection duration led to reduction of the in-cylinder pressure and temperature due to decrease of the injection velocity and evaporation of gasoline spray. They found soot emissions to be highly dependent on the injection duration: with the increase of the injection duration from 6° CA to 18° CA, soot emissions was increased from 0.0037 to 0.6173 g/kWh. However, increasing the fuel temperature led to the enhanced spray atomisation and better combustion process with increase in NO_x and decrease in PM emissions. The increase of the fuel temperature impacted positively on *imep* and engine power.

Another promising approach to ultra-low emissions is PCCI, which relies on late injections and high EGR rates to delay the auto-ignition. PCCI offers lower NO_x

and soot reductions; yet it does not exhibit the higher HC, CO, noise emissions, and the very tight requirements for homogeneity and leanness when compared to HCCI. In a work to determine the impacts of compression ratio on exhaust emissions from a PCCI diesel engine, Laguitton et al. [32] established that, with negligible CO and HC penalty, either decreasing the compression ratio or retarding the injection timing noticeably reduced NO_x and soot emissions when both premixed and diffusion-combustion stages were present. They discovered that lower pressure and temperature during the injection and combustion promoted air-fuel mixing. However, the constraint with PCCI is that it is difficult to achieve at high engine loads. This fact coupled with cold start challenges, and the high emissions of CO and HC make the use of catalysts imperative in PCCI operation [24].

1.5.2 Exhaust Gas Recirculation

The exhaust gas recirculation (EGR) has become a veritable technique in modern diesel engines due to its usefulness in reducing NO_x emissions. For instance, the three-way catalytic converter is not suitable for mitigating NO_x emissions in a diesel engine because the engine works with an excess of air. The technique entails recirculating a certain percentage of the exhaust gas into the inlet air so as to reduce the oxygen content and combustion temperature. This in turn lowers the amount of atmospheric nitrogen that is converted to NO_x. Hence, adding EGR to the inlet mixture reduces the oxygen partial pressure in the inlet mixture, which consequently results in the decline of in-cylinder production of NO_x [42]. In a diesel engine, the inert exhaust gases replace some of the excess oxygen in the pre-combustion mixture. However, in an SI engine about 5–20% of the exhaust gases are fed back into the intake manifold as EGR [43]. It is noteworthy that EGR utilises combustion method that utilises high equivalence ratio because of the extensive use of cold exhaust gas recirculation EGR [31, 39].

The effect of EGR on performance and exhaust emissions has been widely investigated in various engines with promising outcomes. For instance, adding EGR into the intake charge has been found to be the most practical means of controlling charge temperature in an HCCI engine. The effects of cooled EGR on combustion, performance and emission characteristics in HCCI operation region were investigated by Yao et al. [40]. Their findings indicated that while the EGR rate could broaden the HCCI operating region; it had little effect on the maximum load of the HCCI engine fuelled with DME/methanol. In a recent study, it was established by Ghazikhani et al. [36] that EGR could extend the operating range of gasoline-diesel dual-fuel HCCI engine to high equivalence ratios due to reduction of maximum combustion temperature. Furthermore, as found by Nemati et al. [33], increase in EGR rate resulted in reduction in NO_x emissions, which was attributed to the drop in oxygen concentration. However, there was an increase in PM emissions at increased and lower EGR rates; while the drop in in-cylinder pressure occasioned by the increasing EGR rates resulted in power loss.

In one of the most recent work on HCCI, Canakci [44] investigated the effect of inlet air pressure on the combustion characteristics and exhaust emissions of a DI-HCCI gasoline engine under cooled EGR. The study revealed that the mean effective pressures (*mep*) of the engine increased with the increasing boost pressure and decreased with the increasing engine speed. The boost pressure and SOI timing were found to have more effect on the combustion efficiency than the engine speed. Furthermore, increasing the boost pressure resulted in increased brake thermal efficiency and decreased combustion efficiency. The exhaust temperature increased with increasing engine speed for each boost pressure. However, it decreased with the increasing boost pressure since advanced injection timing at higher boost triggered low-temperature reaction. In general, NO_x and CO emissions of the engine decreased when the boosting is employed, while HC emissions increased remarkably.

Conventionally, dual-fuel engines are associated with lower thermal efficiency, higher CO and unburned HC emissions, especially at part load. Hence, EGR is sometimes utilised in partly resolving these problems and attaining further reduction in NO_x emission. Abdelaal and Hegab [34] investigated the effects of variable volumes of partly-cooled EGR (5, 10, and 20%) on the combustion and exhaust emissions of a single-cylinder direct injection (DI) diesel engine, which was modified to run on natural gas as the main fuel and diesel fuel as a pilot fuel. From their findings, the application of EGR in the dual-fuel mode resulted in significant reduction in NO_x emissions. Expectedly, the more the EGR volume, the higher the reduction in NO_x emission. HC and CO emissions at dual-fuel mode were higher than the emissions at conventional diesel mode, especially at engine part loads. Nonetheless, there were reductions in HC and CO emissions at dual-fuel mode with increase in engine load. They, however, stated that the application of EGR to dual-fuel mode somewhat moderated HC and CO emissions, but their values were still significantly higher than that of conventional diesel mode. They reported that at part loads, dual-fuel mode exhibited a lower thermal efficiency than conventional diesel mode. The reverse was, however, the case at high loads where dual-fuel mode produced a higher efficiency. The use of EGR was found to have altered the thermal efficiency either by increasing or decreasing depending on the load conditions. They concluded that dual-fuel mode with EGR demonstrated thermal efficiency similar to conventional diesel mode.

Although it helps to reduce NO_x considerably; nevertheless, EGR in a diesel engine has certain shortcomings. Less effective combustion and increased soot production are noticeable disadvantages of EGR technology. The drop in oxygen concentration and overall temperature favours the release of HC and CO emissions; hence, it mitigates NO_x emissions at the expense of rise in the emissions of HC, CO and PM [45]. Moreover, there is a bound to the extent that EGR technology can reduce NO_x, which is around 35%, necessitating further treatment of exhaust gases in order to keep pace with stringent emissions standards. The effect of increasing EGR in diesel engine on the resultant NO_x and PM emissions was discussed by Hountalas et al. [46]. The large amount of soot in higher EGR percentage was attributed to lower performance due to reduction in oxygen concentration, and soot accumulation due to the recirculation of exhaust gas through many cycles as well.

1.6 Fuel-Based Emissions Techniques

Proper modifications to the types of fuel and its composition hold vast prospects in the search for a long-term solutions to the hazard of engine-out emissions. The economy, flexibility, effectiveness and the simplicity of retrofitting into existing engines make this method unique. Although there are several instances of this method undergoing research and development, Compressed Natural Gas/Diesel dual-fuel technique, simulating EGR through CO_2, Biogas/Diesel dual-fuel, and Liquefied Petroleum Gas (LPG)–Diesel dual-fuel are discussed in the following sections.

1.6.1 Compressed Natural Gas/Diesel (CNG-Diesel) Dual-Fuel Combustion Engines

In retrospect, new opportunities for alternative fuels were triggered by the oil crisis of 1973. The need to devise durable solutions to fluctuating prices of crude oil and stringent emission standards has positioned natural gas one of the promising alternatives to fossil fuels. Natural gas, frequently stored in vessels at 20 MPa, has been extensively acknowledged as one the best alternatives to fossil fuels because of its inherent clean nature of combustion [47]. The carbon mass percentage in natural gas is about 75% compared to 86–88% for both petrol and diesel. Hence, the low carbon-to-hydrogen ratio in natural gas engines results in cleaner combustion and subsequently, lower emissions of CO_2, CO and HC than petrol or diesel engines per unit of energy released [48]. Furthermore, the combustion of natural gas practically produces negligible particulates since natural gas contains less dissolved impurities. Natural gas is, therefore, the cleanest burning fuel available for direct utilisation in IC engines. Moreover, due to the fact that it is introduced into the engine as a gas rather than as a liquid, it often provides quicker cold starts with lower emissions. Hence, there is gradual increase in CNG-Diesel dual-fuel engines due to improved fuel efficiency and the potential for reduction in NO_x and PM emissions, which are really difficult to achieve in diesel engines.

The dual-fuel combustion concept (CNG-Diesel) is a veritable option in the quest towards the moderation of exhaust emissions from diesel engines. A carburetted mixture of air and high octane gaseous fuel (e.g. natural gas), prepared in an external mixing device, is compressed and inducted into the engine cylinders. The compressed mixture of air and natural gas does not auto-ignite due to its high auto-ignition temperature. Hence, a pilot quantity of diesel fuel is injected into the cylinders prior to the top dead centre (TDC) of the compression stroke. After a short ignition delay, the combustion of diesel occurs first; igniting the natural gas, and then the flame propagation begins. The presence of compressed natural gas (CNG) in the charge reduces the cylinder temperature due to lean combustion, which subsequently results in lower NO_x and PM emissions due to the drop in quantity of diesel oil utilised during combustion.

Extensive researches have been conducted to appraise the combustion and emission characteristics of dual-fuel compression ignition engines operated with pilot diesel fuel and natural gas. Such studies have revealed that the dual-fuel concept employing CNG/Diesel is veritable in the mitigation of emissions from diesel engines and contribution to improved fuel economy. Liu et al. [49] investigated the emissions of a CNG/diesel dual-fuel direct injection diesel engine with different pilot fuel quantities and optimised pilot injection timing. Their findings revealed that the effects of pilot fuel quantity and injection timing were evident and significant for CNG/diesel dual-fuel engines. They reported that under dual-fuel mode, CO emissions were noticeably higher than the emissions under normal diesel operation mode even at high load, which were triggered by the flame quenching of the lean premixed natural gas–air mixture. Under dual-fuel mode NO_x emissions were reduced by approximately 30% in comparison to diesel mode. They attributed the reduction in NO_x emissions to the fact most of the fuel was burned under lean premixed conditions which resulted in lower local temperature. There were reductions in PM emissions under dual fuelling mode due to the premixed combustion and the molecular structure of methane. However, as pilot fuel quantity was increased, PM emission also increased. They further reported that the unburned HC emissions under dual-fuel mode were evidently higher than that of the normal diesel mode, particularly at low and medium loads. However, significant reduction in HC emissions were achieved with increase in the pilot diesel quantity.

Singh and Maji [50] studied the effect of compression ratio effect on the performance of a dual-fuel diesel engine. They found that enhanced compression ratio positively impacted on the reduction of gaseous and particulate emissions, especially when the dual-fuel engine operated at high percentage of CNG substitution (as high as 60%) and rated load. In this study, up to 85 and 93% smoke opacity reductions were achieved. In another study, Çelıkten [51] experimentally investigated the effects of injection pressure on engine performance and exhaust emissions. They reported that by varying diesel injection pressure, lower NO_x, CO and HC emissions were recorded. The lower NO_x, CO and HC emissions were attributed to the jet penetration and combustion spread in the chamber. Based on their findings, they recommended high injection pressure for keeping O_2, SO_2, and CO_2 at low levels; and low injection pressure to enhancing reduction in NO_x emissions and low level of exhaust smoke.

Furthermore, Gharehghani et al. [52] reported that a reduction in NO_x emission can be achieved by increasing the intake swirl ratio in dual-fuel engines, perhaps due to the increase in heat losses as swirl increases. The study revealed significant reduction in NO_x at high engine load with lower mass ratio of natural gas, while lower PM was recorded at lower engine load and higher mass ratio of natural gas. At part load, they discovered that, the total BSFC was considerably higher under dual-fuel operation mode due to the low combustion rate of gaseous fuel. Similar trend in performance was observed by Saray and Pirouzpanah [53] with higher loads of the dual-fuel combustion operation. However, at lower loads both emission and combustion suffered due to the extremely lean mixture and incomplete combustion.

1.6.2   Simulating EGR Through CO_2

In furtherance of the sustained efforts by researchers in mitigating NO_x and PM emissions, moderate attempts have been made at unravelling the feasibility of using CO_2 in simulating the roles of EGR. In one of the pioneering work, Zhao et al. [54] utilised CO_2 in simulating the effects of EGR. The findings indicated that CO_2 addition was found suitable in mitigating NO_x and soot emissions; and the cost of EGR cooling system as well. In a much later study, Jin et al. [55] used four different CO_2 mass fractions in diesel fuel such as 3.13, 7.18, 12.33 and 17.82% to study the effect of CO_2 concentration on jet flame characteristics. They reported that the dilution and atomisation effects of the dissolved CO_2 component in the fuel had a great influence on the flame structure. At the same injected pressure, the low temperature flame length increased with the increase of CO_2 concentration in the fuel while the total flame length and high temperature flame length decreased with the increase of CO_2 concentration in the fuel. The mean temperature of flame increased and then dropped with the increase of CO_2 concentration in the fuel at the same injected pressure.

The effects of CO_2 in-cylinder injection on PCCI combustion were investigated by Qu et al. [56]. Their findings revealed that, for CO_2 in-cylinder injection, either advancing CO_2 injection timing or increasing cyclic injection quantity greatly reduced the emission of NO_x with little impairment of the thermal efficiency. With increase in the CO_2 injection quantity, intake pipe injection had a similar effect to EGR, such as reduction in NO_x emission accompanied by increase in PM. While the CO_2 in-cylinder injection had a weaker effect in reducing NO_x emission than the intake-pipe injection, it had a little effect on smoke [56].

Furthermore, Çinar et al. [57] investigated the effect of using CO_2 as a diluent in the intake manifold and injection pressure on engine performance and emissions. In this study, CO_2, used as diluent, was introduced to the intake manifold of a diesel engine at a ratio of 2, 4 and 6% respectively. NO_x emission was found to be high for moderate injection pressures at low concentrations of CO_2. They reported that as the intake of CO_2 increased, NO_x emission considerably decreased. They attributed the drastic reduction in NO_x emission to the fact that molar heat capacity of CO_2 is higher than that of atmospheric air. Hence, CO_2 absorbs more combustion enthalpy thereby considerably reducing in-cylinder temperature and NO_x formation rate. Their findings further revealed that while NO_x emission reduced with CO_2 admission in the inlet charge, other parameters deteriorated. With 6% CO_2 admission, engine torque, power, *bmep* and *bsfc* approximately deteriorated by 5.9, 5.5, 6, and 3.3% respectively; smoke emission increased approximately by 60%, while CO emission increased approximately by 8.5 times from its base level. In spite of all these, NO_x emission reduced approximately by 50% at 6% admission of CO_2.

Bedoya et al. 2009 [58] investigated the effects of mixing system and the pilot fuel quality on performance using biogas (60% CH_4 + 40% CO_2) as primary fuel in a stationary dual-fuel, four-stroke, two-cylinder DI power generation diesel engine. Their findings revealed that full diesel substitution was achievable using biogas and biodiesel as power sources for all loads evaluated. Thermal efficiency and substitution of pilot fuel were increased, while CH_4 and CO emissions were reduced by using the supercharged mixing system combined with biodiesel as pilot fuel.

1.6.3 Biogas/Diesel Dual-Fuel

In Duc and Wattanavichien [59] the effect of a small IDI biogas premixed charge diesel dual-fuelled CI engine, used in agricultural applications, on combustion and emissions were investigated by varying engine loads and EGR percentages. They established that biogas-diesel dual-fuelling of HCCI engine showed almost no deterioration in engine performance at all test speeds. Lower energy conversion efficiency was recorded, which was offset by the reduced fuel cost of biogas over diesel. In a related work, the prospects of effective utilisation of biogas in HCCI engine using a single-cylinder diesel engine modified to run in HCCI mode was examined by Nathan et al. [60]. The varied parameters were charge temperature and diesel injection quantity. He reported that the CO_2 in the biogas suppressed the high heat release (HHR) prevalent in HCCI engines fuelled with diesel. Thermal efficiencies close to diesel engine operation along with extremely low levels of NO_x ($\leq$20 ppm) and smoke ($\leq$0.1 BSU) were attained in a BMEP range of 2.5 to 4 bar. HC emissions were very high but were lowered when the charge temperature was raised. The thermal efficiency at a BMEP of 4 bar was 27.2% in the biogas-diesel HCCI mode as against 30% with diesel operation.

Makareviciene et al. [61] evaluated the impact of the CO_2 concentration in biogas on the performance and exhaust emissions of a diesel engine, with EGR system installed, running on a mixture of biogas and diesel as pilot fuel. Expectedly, they reported lower pollutant levels when the engine was operated with the EGR system. However, according to their finding the decrease in NO_x emissions was directly proportional to the concentration of methane in the common fuel mixture. Furthermore, they employed the gas with the highest methane content in determining the impact of the start of injection timing on the engine operating parameters. They observed that as the methane content in the common fuel mixture increased, the start of injection timing had to be progressively advanced to increase the thermal efficiency and lower the fuel consumption, the CO and HC emissions and the exhaust smoke. Hence, if the gas is supplied without the engine control system for SOI timing, the total fuel consumption would increase as the thermal efficiency decreases, while the concentration of the exhaust emissions would increase; the only exception being NO_x emissions. However, advancing the start of injection (SOI) timing resulted in higher NO_x emissions. The combined use of EGR system and the control of start of injection timing were, therefore, found to bear positive impacts on performance and emissions.

They reported that when the EGR system was turned off, the air/fuel ratio increased while smoke, CO and HC emissions were reduced. However, NO_x concentration increased while the excessively high air/fuel ratio resulted in higher fuel consumption and lower thermal efficiency. They concluded in quantitative term that brake thermal efficiency comparable to that of an engine running on diesel fuel was achievable by injecting a gas containing about 30% methane (M95% at 30 l/min) to a diesel engine running at an engine speed of 2500 min^{-1} at engine load of 60 Nm (a mean effective pressure of 0.6 MPa). This was attained for the same smoke emissions; thus, reducing the diesel fuel consumption by 1.5 times, and the NO_x emissions by 1.5 times for negligible rises in the CO and HC exhaust emissions.

1.6.4   Liquefied Petroleum Gas (LPG)–Diesel Dual-Fuel

In a study by Qi et al. [62], the effects of LPG–Diesel blended fuel on the performance and pollutant emissions of a DI diesel engine were investigated. They found that upsurge in LPG by mass fraction in the blended fuel resulted in lower peak cylinder pressure. Contrary to diesel operation, at low engine load, equivalent *bsfc* deteriorated under blended fuel mode. However, at high load, equivalent *bsfc* values were close to the values recorded under diesel operation. They reported that there was a drop in NO_x emissions as the mass fraction of the LPG fraction in the blended fuel increased; yielding lower NO_x emissions when compared to diesel operation. Moreover, at high engine load there was a significant reduction in CO emissions as the mass fraction of the LPG fraction increased. However, a slight increase in CO emissions was recorded at low engine load. Emissions of HC marginally increased with upsurge in the mass fraction of the LPG fraction in the blended fuel. Hence, LPG-Diesel blended fuel is a promising technique for controlling engine-out emissions especially NO_x and smoke. This is particularly significant in view of the teething troubles of controlling NO_x and smoke emissions in diesel engines. The drawback in equivalent *bsfc* is partly traded-off by the relative affordability of LPG.

Elnajjar et al. [63] experimentally investigated the effect of different blends of LPG fuel and engine parameters on the performance of a dual-fuel engine running on diesel as a pilot fuel. The mixtures of LPG fuels were Propane to Butane with volume ratio of 100:0, 70:30, 55:45, 25:75, and 0:100. They reported that the different blends of LPG fuel had direct effect on the levels of combustion noise with minor effect on the engine efficiency with fuel with the volume ratio 100:0 exhibiting the highest noise levels. Advancing the pilot fuel injection timing resulted in a drop in the overall efficiency. They discovered that the volume ratio 25:75 was the best in terms of performance, achieving the highest level of efficiency and lowest noise levels when compared with the other fuel types. The lowest efficiency with the highest levels of noise was exhibited by the volume ratio 55:45.

Furthermore, as mass flow rates of the gaseous fuel were increased, all fuel types tended to have higher efficiency accompanied with higher engine combustion noise. The highest rate of pressure rise was recorded by fuel with the volume ratio 100:0. On the other hand, increasing the mass flow rate of pilot fuel yielded higher engine overall efficiency while the combustion noise levels were slightly reduced. They concluded that the measured engine thermal efficiency for all the blends were similar; however, they were differences in the levels of noise generated by the engine. The fuel with the volume ratio 100:0 created the lowest levels of noise with comparable engine performance to others under all varying parameters.

In a related work, Saleh [64] conducted a study to examine the effects of five gaseous fuel with different compositions of LPG on performance and exhaust emissions in a dual fuel engine operated on diesel as pilot fuel. Their findings showed that the exhaust emissions, the exhaust gas temperatures and fuel conversion efficiency of the dual-fuel engine were affected when different LPG compositions were used, with higher butane content leading to lower NO_x levels while higher propane content resulted in reduced CO levels. Based on their findings, they suggested that a better trade-off between CO and NO_x emissions could be attained within EGR rate of 5–15%. NO_x and SO_2 emissions for fuel #3-diesel blend decreased by 27%, 69% at full load and 35%, 51% at 25% load respectively, in comparison with the conventional diesel engine. However, CO emissions increased about 15.7% at full load and deteriorated up to 100% at 25% load.

Moreover, they reported that fuel #3 (70% propane, 30% butane) with mass fraction 40% replacement of the diesel fuel was the best LPG composition in the dual-fuel operation because the overall engine performance was equivalent to the conventional diesel engine except at part loads. With EGR rate at 5%, the fuel conversion efficiency of fuel #3-diesel blend was improved at part loads. Hence, they concluded that test engine was capable of smoothly running up to 90% of diesel fuel substitution and the ratio of $M_{propane}/(M_{diesel} + M_{propane}) = 40\%$ was the best for maintaining the high thermal efficiency comparable to a conventional engine.

1.7 Fuel Injection Method

The port fuel injection (PFI) or multi-point fuel injection (MFI) method in metering gases into the engine is still widely in use today [65]. It is the most efficient method of metering gases into IC engines. MFI injects fuel into the intake ports just upstream of the intake valve or valves, rather than spraying the fuel at a central point inside an intake manifold. Injecting the fuel this close to the intake port almost ensures that it will be drawn completely into the cylinder. MFI can be sequential, in which fuel injection is programmed to coincide with each cylinder's intake stroke [26, 27]. It can also be the batched type, which injects fuel into the cylinders in groups, without precisely harmonising any specific cylinder's intake stroke. The third instance is simultaneous, in which fuel is introduced into all the cylinders at the same time. The main advantage of MFI is that it meters fuel more precisely, achieving the desired air/fuel ratio while improving all related aspects. Also, it virtually eliminates the possibility that fuel will condense or collect in the intake manifold. It is noteworthy that many modern electronic fuel injection (EFI) systems utilise the sequential type of MFI. Nevertheless, direct injection systems are now replacing the sequential types in modern gasoline engines.

In dual-fuel (gas–liquid diesel) applications, the fumigation method is often utilised in introducing gaseous fuels into the intake ports. The method entails blending a gaseous fuel with incoming air in correct proportion before entering the intake ports together. The gaseous fuel is not the primary fuel in a dual-fuel engine, hence the fumigation method works perfectly well. Moreover, no engine modification is required in utilising the technique. Whereas the direct injection of gaseous fuel is expensive and complex, the fumigation method is not. According to Papagiannakis [66], the amount of the gaseous fuel fumigated into the intake air displaces an equal amount of the inducted combustion air since at constant engine speed the total amount of the inducted mixture must be kept constant, while the desired engine power output, that is the engine load, is controlled by changing the amount of the natural gas.

Adhering to experimental procedure, while operating under dual-fuel mode at a specified engine speed, sufficient quantity of diesel oil is injected to achieve a percentage of the desired engine power output. Thereafter, the remaining percentage of the desired engine power output is attained by using a gaseous fuel, for instance natural gas, which is fumigated into the air intake. The liquid diesel fuel supplement ratio x, which denotes the quotient of the mass flow rate of natural gas divided by the total fuel (diesel and natural gas) mass flow rates as reported by Papagiannakis et al. [30] is given by the expression:

$$x = \frac{\dot{m}_{NG}}{\dot{m}_D + \dot{m}_{NG}} \cdot 100 \, (\%) \tag{1.1}$$

The term $\dot{m}_D$ denotes diesel oil consumption usually measured by a flow meter suitable for liquid fuel, while $\dot{m}_{NG}$ is the gaseous fuel consumption measured by a rotameter or a mass flow meter. Moreover, they expressed the corresponding total relative air–fuel ratio for both fuels as:

$$\lambda_{total} = \frac{\dot{m}_{air}}{AFR_{NG}^{st} \cdot \dot{m}_{NG} + AFR_D^{st} \cdot \dot{m}_D} \tag{1.2}$$

where AFR_D^{st} and AFR_{NG}^{st} represent the stoichiometric air–fuel ratio (by mass) for the diesel fuel and the natural gas, respectively. Moreover, the desired engine load (i.e. *bmep*) is attained by changing the amounts of the injected fuels. Liquid diesel fuel is utilised to attain a percentage of the desired engine load, while the remaining the fraction of the desired *bmep* is achieved by fumigating natural gas into the intake air.

Using timed port injection (TPI) technique, Lakshmanan and Nagarajan [67] investigated the performance and emission characteristics of a DI diesel engine running on acetylene gas with diesel as the source of ignition. Acetylene was fumigated into the intake port as a secondary fuel while diesel was directly injected into the cylinder. The gas flow rates were 110, 180 and 240 g/h, while the engine loads were varied from low to full load. They reported that the optimum conditions were 5° ATDC injection timing and 90° CA injection duration. Brake thermal efficiency for the dual fuel mode was found to be very close to diesel operation mode at full load for the maximum gas flow rate. They recorded reductions in NO_x, HC and CO emissions when compared to diesel operation, while there was negligible increase in smoke emission. NO_x emission reduced by 15% at 110 g/h, 11% at 180 g/h, and 9% at 240 g/h of gas flow rates when compared to diesel fuel operation at full load. On the other hand, smoke level slightly increased by 5.26% at 240 g/h, 8.52% at 180 g/h, and 12.78% at 110 g/h of gas flow rates at full load when compared to baseline diesel operation. They reported that CO_2 emission decreased by 5% at 110 g/h, 11% at 180 g/h, and 18% at 240 g/h of gas flow rate at full load compared to base line diesel operation.

Obviously, the unique characteristics of acetylene such as higher burning velocity, leaner operating capability, and higher energy content aided complete combustion of the cylinder charge, which in turn resulted in reduction in HC emissions. They discovered that by varying acetylene flow rate from 110 to 240 g/h, it was feasible to smoothly operate a DI diesel engine with stable combustion using acetylene as a fuel and diesel as pilot ignition. Hence, they concluded that timed port injection of acetylene in dual-fuel mode resulted in safe operation of the engine with reduction in NO_x, HC, CO, and CO_2 emissions, with no depreciation in thermal efficiency.

1.8 Summary

Undoubtedly, meeting future emissions regulations is one of the biggest challenges that stakeholders in the automotive industry have ever faced. Therefore, researchers have been unrelenting in terms of research and development in finding globally acceptable solutions, which are cost-effective driven, to the menace of exhaust emissions of automotive engines. Modern hardware-based techniques still enjoy prominence; however, they are prone to salient setbacks such as limited durability, high cost and difficulty in installing on existing engines, among others. Investigation has, therefore, revealed that there are no one-size-fits-all solutions in place. Most of the existing solutions enjoy, at best, relative advantages over one another.

No matter how one views the trend of events, what is obvious is that for now, high cost is inevitable in knocking out exhaust emissions of automobiles while sustaining power and performance. It is, therefore, safe to suggest that the enduring solutions in vogue, make the eventual costs of automobiles expensive; and sometimes beyond the reach of the masses. In fact, manufacturers, in particular, are accepting increased costs as a result of research and development, essentially in areas of emissions reduction. Consequently, the overall costs of engines have increased due to components improvement for healthy emissions.

The significant revelation of this study were the plethora of prospects that the fuel-based and combustion-based techniques could offer stakeholders in the pursuit of enduring panacea to the menace of engine-out emissions. In most of the scholarly work that were examined, the two techniques brought about significant reductions in exhaust emissions. However, it must be stated that in some of the cases reviewed, while deterioration in HC emissions were recorded, there were no significant improvements in the trade-off relationship between PM and NO_x emissions.

Hence, there are openings for research and development in these areas in the quest towards finding sustainable solutions to the threats that exhaust emissions from automotive engines pose to humans and the environments.

Expectedly, based on chemistry of combustion, the potential of the dual-fuel approach through the utilisation of natural gas as base fuel in diesel engine in simultaneously mitigating PM and NO_x exhaust emissions were duly established. Perhaps, the most unique fact that emerged from this work is the discovery that, where noticeable reductions in emissions were attained, carbon dioxide (CO_2) was central to most of the techniques utilised in those instances. This is, no doubt, a significant finding. Hence, CNG-CO_2/Diesel dual-fuel utilisation in diesel engine should be more promising in the extenuation of engine exhaust emissions. The technique aims at using CNG-Diesel dual-fuel combustion with high concentration of CO_2 in the CNG. Usually, natural gas contains 4 to 50% of CO_2 gas [68] and in some extreme situations 28 to 87% [69]. The CO_2 weakens oxygen concentration and reduces temperature through its higher heat absorbing capacity relative to the air it displaces. This ensures the utilisation of CNG with high CO_2 as automotive fuel, which avoids the expensive separation processes of the two gases. The presence of CO_2 in the CNG will, typically, have similar effects as EGR.

Markedly, inferences from cases studied, revealed that existing literature on combustion, performance and emission characteristics of CNG-CO_2/Diesel dual-fuel utilisation in diesel engine is still scanty. This explains why available data on in-cylinder pressure tracing and temperature; rate of heat release and other performance parameters characterising CNG-CO_2/Diesel combustion are still inadequate. It is equally noteworthy that data on the viability or otherwise of improvement in engine performance parameters through an optimised mix of operation parameters are not available. Hence, further scholarly investigations in the nearest future, comprehensive in all ramifications, will be necessary. This concept should find applications in engines which do not have EGR system to mitigate exhaust emissions, with enhanced fuel efficiency compared to the conventional diesel engine without the need for major modifications.

References

1. P.F. Nelson, A.R. Tibbett, S.J. Day, Effects of vehicle type and fuel quality on real world toxic emissions from diesel vehicles. Atmos. Environ. **42**, 5291–5303 (2008)
2. A.K. Agarwal, Biofuels (alcohols and biodiesel) applications as fuels for internal combustion engines. Prog. Energy Combust. Sci. **33**, 233–271 (2007)
3. S. Pinzi, I. Garcia, F. Lopez-Gimenez, M. Luque de Castro, G. Dorado, M. Dorado, The ideal vegetable oil-based biodiesel composition: a review of social, economical and technical implications. Energy Fuels **23**, 2325–2341 (2009)
4. C.L. Cummins, Diesels for the first stealth weapon: submarine power 1902–1945. in *US Naval Institute Proceedings* (2009), p. 74
5. J. Heywood, *Internal Combustion Engine Fundamentals* (McGraw-Hill Education, 1988)
6. O. Laguitton, Advanced diesel combustion strategies for ultra-low emissions, University of Brighton (2005)

7. Integer Research Limited, Passenger cars and light duty vehicles, Emissions Legislation (2012), Available: http://www.integer-research.com/legislation/europe-russia/. Accessed: 17 Oct 2016 [Online]
8. A. Mamakos, C. Dardiotis, A. Marotta, G. Martini, U. Manfredi, R. Colombo et al., in *Particle Emissions from a Euro 5a Certified Diesel Passenger Car*. JRC Scientific and Technical reports, European Union (2011)
9. Regulation (EC) No 715/2007 of the European Parliament and of the Council of 20 June 2007. Off. J. Eur. Union **50**(L 171), 1–16 (2007)
10. Regulation (EC) No 692/2008 of 18 July 2008 implementing and amending Regulation (EC) No 715/2007 of the European Parliament and of the Council. Off. J. Eur. Union **51**(L 199), 1–135 (2008)
11. Regulation (EU) 2016/427 of 10 March 2016 amending Regulation (EC) No 692/2008. Off. J. Eur. Union **59**(L 82), 1–98 (2016)
12. Regulation (EC) No 443/2009 of the European Parliament and of the Council of 23 April 2009. Off. J. Eur. Union **52**(L 140), 1–15 (2009)
13. Regulation (EU) No 510/2011 of the European Parliament and of the Council of 11 May 2011. Off. J. Eur. Union **54**(L 145), 1–18 (2011)
14. Communication from the Commission to the Council and the European Parliament on the Strategy for reducing heavy-duty vehicles' fuel consumption and CO_2 emissions (The European Commission, Brussels, Belgium, 2014)
15. K. Lindqvist, Emission standards for light and heavy road vehicles. AirClim Factsheet (2012)
16. Regulation (EU) No 253/2014 of the European Parliament and of the Council of 26 February 2014 amending Regulation (EU) No 510/2011. Off. J. Eur. Union **57**(L 84), 38–41 (2014)
17. Regulation (EU) No 333/2014 of the European Parliament and of the Council of 11 March 2014 amending Regulation (EC) No 443/2009. Off. J. Eur. Union **57**(L 103), 15–21 (2014)
18. China: Beijing launches Euro 4 standards. Automotive World (2008), Available at: http://www.automotiveworld.com/. Accessed: 17 Oct 2016 [Online]
19. H.K. Suh, S.W. Park, C.S. Lee, Effect of piezo-driven injection system on the macroscopic and microscopic atomization characteristics of diesel fuel spray. Fuel **86**, 2833–2845 (2007)
20. A. Ibrahim, S. Bari, Effect of varying compression ratio on a natural gas SI engine performance in the presence of EGR. Energy Fuels **23**, 4949–4956 (2009)
21. J.E. Dec, Advanced compression-ignition engines—understanding the in-cylinder processes. Proc. Combust. Inst. **32**, 2727–2742 (2009)
22. I.D. Bedoya, S. Saxena, F.J. Cadavid, R.W. Dibble, M. Wissink, Experimental study of biogas combustion in an HCCI engine for power generation with high indicated efficiency and ultra-low NO_x emissions. Energy Convers. Manag. **53**, 154–162 (2012)
23. M. Yao, Z. Zheng, H. Liu, Progress and recent trends in homogeneous charge compression ignition (HCCI) engines. Prog. Energy Combust. Sci. **35**, 398–437 (2009)
24. J.E. Parks, V. Prikhodko, J.M. Storey, T.L. Barone, S.A. Lewis, M.D. Kass et al., Emissions from premixed charge compression ignition (PCCI) combustion and affect on emission control devices. Catal. Today **151**, 278–284 (2010)
25. S. Gan, H.K. Ng, K.M. Pang, Homogeneous charge compression ignition (HCCI) combustion: implementation and effects on pollutants in direct injection diesel engines. Appl. Energy **88**, 559–567 (2011)
26. B. Pundir, IC engines: combustion and emissions. Alpha Sci. Int. (2010)
27. W. W. Pulkrabek, in *Engineering Fundamentals of the Internal Combustion Engine* (Prentice Hall, 2004)
28. A. Garvine, 'One Giant Leap', interview of Armstrong. Engine Technol. Int. 20–23 (2002)
29. L. Wei, P. Geng, A review on natural gas/diesel dual fuel combustion, emissions and performance. Fuel Process. Technol. **142**, 264–278 (2016)
30. R.G. Papagiannakis, C.D. Rakopoulos, D.T. Hountalas, D.C. Rakopoulos, Emission characteristics of high speed, dual fuel, compression ignition engine operating in a wide range of natural gas/diesel fuel proportions. Fuel **89**, 1397–1406 (2010)

31. H. Machrafi, S. Cavadias, J. Amouroux, Influence of fuel type, dilution and equivalence ratio on the emission reduction from the auto-ignition in an homogeneous charge compression ignition engine. Energy **35**, 1829–1838 (2010)
32. O. Laguitton, C. Crua, T. Cowell, M. Heikal, M. Gold, The effect of compression ratio on exhaust emissions from a PCCI diesel engine. Energy Convers. Manag. **48**, 2918–2924 (2007)
33. A. Nemati, R. Barzegar, A.S. Khalil, H. Khatamnezhad, Decreasing the emissions of a partially premixed gasoline fueled compression ignition engine by means of injection characteristics and EGR. Therm. Sci. **15**, 939–952 (2011)
34. M.M. Abdelaal, A.H. Hegab, Combustion and emission characteristics of a natural gas-fueled diesel engine with EGR. Energy Convers. Manag. **64**, 301–312 (2012)
35. M. Torres Garcia, R. Chacartegui Ramirez, F. Jimenez-Espadafor Aguilar, T. Sanchez Lencero, Analysis of the start of combustion of a diesel fuel in a HCCI process through an integral chemical kinetic model and experimentation. Energy Fuels **22**, 987–995 (2008)
36. M. Ghazikhani, M. Kalateh, Y. Toroghi, M. Dehnavi, An experimental study on the effect of premixed and equivalence ratios on CO and HC emissions of dual fuel HCCI engine. Int. J. Mech. Syst. Sci. Eng. **1** (2009)
37. K. Kobayashi, T. Sako, Y. Sakaguchi, S. Morimoto, S. Kanematsu, K. Suzuki et al., Development of HCCI natural gas engines. J. Nat. Gas Sci. Eng. **3**, 651–656 (2011)
38. R. Osborne, G. Li, S. Sapsford, J. Stokes, T. Lake, M. Heikal, in *Evaluation of HCCI for Future Gasoline Powertrains*. SAE Technical Paper 0148–7191 (2003)
39. S. Khalilarya, S. Jafarmadar, H. Khatamnezhad, G. Javadirad, M. Pourfallah, Simultaneously Reduction of NO_x and Soot Emissions in a DI heavy duty diesel engine operating at high cooled EGR rates. Int. J. Aerosp. Mech. Eng. **6**, 1020–1028 (2012)
40. M. Yao, Z. Chen, Z. Zheng, B. Zhang, Y. Xing, Study on the controlling strategies of homogeneous charge compression ignition combustion with fuel of dimethyl ether and methanol. Fuel **85**, 2046–2056 (2006)
41. K. Duffy, A. Kieser, E. Fluga, D. Milam, in *Heavy-Duty HCCI Development Activities*. 2004 DEER Conference, Department of Energy, EERE, Coronado, CA, USA (2004)
42. G. Abd-Alla, H. Soliman, O. Badr, M. Abd-Rabbo, Effects of diluent admissions and intake air temperature in exhaust gas recirculation on the emissions of an indirect injection dual fuel engine. Energy Convers. Manag. **42**, 1033–1045 (2001)
43. A.K. Sen, S.K. Ash, B. Huang, Z. Huang, Effect of exhaust gas recirculation on the cycle-to-cycle variations in a natural gas spark ignition engine. Appl. Therm. Eng. **31**, 2247–2253 (2011)
44. M. Canakci, Combustion characteristics of a DI-HCCI gasoline engine running at different boost pressures. Fuel **96**, 546–555 (2012)
45. V. Peixoto, C. Argachoy, I. Trindade, and M. Airoldi, Combustion optimization of a diesel engine with EGR system using 1D and 3D simulation tools, in *The Fourth European Combustion Meeting*, Department of Engine Design Engineering, MWM International Diesel Engines of South America Ltd. Sao Paulo, Brazil (2009)
46. D. Hountalas, G. Mavropoulos, K. Binder, Effect of exhaust gas recirculation (EGR) temperature for various EGR rates on heavy duty DI diesel engine performance and emissions. Energy **33**, 272–283 (2008)
47. W. Abdelghaffar, Performance and emissions of a diesel engine converted to dual diesel–CNG fuelling. Eur. J. Sci. Res. **56**, 279–293 (2011)
48. L. Turrio-Baldassarri, C.L. Battistelli, L. Conti, R. Crebelli, B. De Berardis, A.L. Iamiceli et al., Evaluation of emission toxicity of urban bus engines: compressed natural gas and comparison with liquid fuels. Sci. Total Environ. **355**, 64–77 (2006)
49. J. Liu, F.Y. Yang, H.W. Wang, M.G. Ouyang, S.G. Hao, Effects of pilot fuel quantity on the emissions characteristics of a CNG/diesel dual fuel engine with optimized pilot injection timing. Appl. Energy **110**, 201–206 (2013)
50. R. Singh, S. Maji, Performance and exhaust gas emissions analysis of direct injection cng-diesel dual fuel engine. Int. J. Eng. Sci. Technol. **4**, 833–846 (2012)

51. I. Çelıkten, An experimental investigation of the effect of the injection pressure on engine performance and exhaust emission in indirect injection diesel engines. Appl. Therm. Eng. **23**, 2051–2060 (2003)
52. A. Gharehghani, S. Mirsalim, S. Jazayeri, Numerical and experimental investigation of combustion and knock in a dual fuel gas/diesel compression ignition engine. J. Combust. **2012** (2012)
53. R.K. Saray, V. Pirouzpanah, Theoretical investigation of combustion process in dual fuel engines at part load considering the effect of exhaust gas recirculation. J. Engine Res. **14**, 51–62 (2009)
54. H. Zhao, J. Hu, N. Ladommatos, In-cylinder studies of the effects of CO_2 in exhaust gas recirculation on diesel combustion and emissions, in *Proceedings of the Institution of Mechanical Engineers, Part D: Journal of Automobile Engineering*, vol. 214 (2000), pp. 405–419
55. X. Jin, H. Zhen, Q. Xinqi, H. Yuchun, The effect of CO_2 dissolved in a diesel fuel on the jet flame characteristics. Fuel **87**, 395–404 (2008)
56. S. Qu, K. Deng, Y. Cui, L. Shi, Effects of carbon dioxide in-cylinder injection on premixed charge compression ignition combustion, in *Proceedings of the Institution of Mechanical Engineers, Part D: Journal of Automobile Engineering*, vol. 222 (2008), pp. 1501–1511
57. C. Çinar, T. Topgül, M. Ciniviz, C. Haşimoğlu, Effects of injection pressure and intake CO_2 concentration on performance and emission parameters of an IDI turbocharged diesel engine. Appl. Therm. Eng. **25**, 1854–1862 (2005)
58. I.D. Bedoya, A.A. Arrieta, F.J. Cadavid, Effects of mixing system and pilot fuel quality on diesel–biogas dual fuel engine performance. Biores. Technol. **100**, 6624–6629 (2009)
59. P.M. Duc, K. Wattanavichien, Study on biogas premixed charge diesel dual fuelled engine. Energy Convers. Manag. **48**, 2286–2308 (2007)
60. S.S. Nathan, J. Mallikarjuna, A. Ramesh, An experimental study of the biogas–diesel HCCI mode of engine operation. Energy Convers. Manag. **51**, 1347–1353 (2010)
61. V. Makareviciene, E. Sendzikiene, S. Pukalskas, A. Rimkus, R. Vegneris, Performance and emission characteristics of biogas used in diesel engine operation. Energy Convers. Manag. **75**, 224–233 (2013)
62. D. Qi, Y.Z. Bian, Z.Y. Ma, C.H. Zhang, S.Q. Liu, Combustion and exhaust emission characteristics of a compression ignition engine using liquefied petroleum gas–diesel blended fuel. Energy Convers. Manag. **48**, 500–509 (2007)
63. E. Elnajjar, M.Y. Selim, M.O. Hamdan, Experimental study of dual fuel engine performance using variable LPG composition and engine parameters. Energy Convers. Manag. **76**, 32–42 (2013)
64. H. Saleh, Effect of variation in LPG composition on emissions and performance in a dual fuel diesel engine. Fuel **87**, 3031–3039 (2008)
65. C. Cheung, Z. Zhang, T. Chan, C. Yao, Investigation on the effect of port-injected methanol on the performance and emissions of a diesel engine at different engine speeds. Energy Fuels **23**, 5684–5694 (2009)
66. R. Papagiannakis, Comparative Evaluation of the effect of partial substitution of diesel fuel by natural gas on performance and emissions of a fumigated dual fuel diesel engine. Gas **3**, 6 (2011)
67. T. Lakshmanan, G. Nagarajan, Experimental investigation of port injection of acetylene in DI diesel engine in dual fuel mode. Fuel **90**, 2571–2577 (2011)
68. A.K. Datta, P.K. Sen, Optimization of membrane unit for removing carbon dioxide from natural gas. J. Membr. Sci. **283**, 291–300 (2006)
69. N. Darman, A. Harun, in *Technical Challenges and Solutions on Natural Gas Development in Malaysia*. Beijing, China, Petronas/Petronas Carigali (2006)

Chapter 2
Rubber Seed/Palm Oil Biodiesel

Ibrahim Khalil Adam, A. Rashid A. Aziz, Morgan R. Heikal, Suzana Yusup and Firmansyah

2.1 Introduction

The use of a clean fuel such as biodiesel has both environmental and economic advantages. The environmental advantage is reduced emissions, whereas the economic benefit stems from local availability of raw material and improved energy security. Currently most of biodiesel production comes from edible sources such as soybean, palm and sunflower oils. However due to price fluctuations, land limitation, contrary to the current social movement and energy policies, their industrial expansion has been limited [1]. Among the edible biodiesel sources palm is the most suitable one as it has a higher oil yield compared to other oil crops [2]. Although the biodiesel production from these sources is inevitable for their availability and

I. K. Adam (✉) · A. R. A. Aziz
Mechanical Engineering Department, Centre for Automotive Research and Electric Mobility, Universiti Teknologi PETRONAS, 32610 Seri Iskandar, Perak, Malaysia
e-mail: himakh80@gmail.com; ibrahim_g02238@utp.edu.my

A. R. A. Aziz
e-mail: rashid@utp.edu.my

M. R. Heikal
School of Computing, Engineering and Mathematics, Advanced Engineering Centre, University of Brighton, Brighton BN2 4GJ, UK
e-mail: m.r.heikal@brighton.ac.uk

S. Yusup
Chemical Engineering Department, Universiti Teknologi PETRONAS, 32610 Seri Iskandar, Perak, Malaysia
e-mail: drsuzana_yusuf@utp.edu.my

Firmansyah
Centre for Automotive Research and Electric Mobility, Universiti Teknologi PETRONAS, 32610 Seri Iskandar, Perak, Malaysia
e-mail: firmansyah@utp.edu.my

Z. A. Abdul Karim and S. A. Sulaiman (eds.), *Alternative Fuels for Compression Ignition Engines*, SpringerBriefs in Energy, https://doi.org/10.1007/978-981-10-7754-8_2

large production levels, reducing their amounts using non-edible sources will relieve them for other uses. Blending edible/non-edible oils is a solution that will have significant contribution towards the advancement of the industry. Jatropha-palm oil [3], Jatropha-soapnut [4] and Mahua-simarouba [5] oil blends had been investigated and observed to be a good potential sources for biodiesel productions. In Malaysia there are 1,229,940 hectares of rubber plantation according to the association of Natural Rubber Producing Countries and the projected annual production is estimated to be 1.2 million metric ton per year [6]. Each tree yields an average of 800 seed (1.3 kg) twice a year depending on crop density and soil fertility. The kernel has an average oil of 40–50 wt% and can be used for biodiesel synthesis [2]. Comprehensive literature on the rubber seed oil based biodiesel production process is available in [7]. Recently, the palm oil based biodiesel usages in a diesel engine have been studied by many researchers [8, 9]. The results indicated that, engine performance were decreased as blend ratio increases and the combustion characteristics slightly changed with biodiesel fuel compared to diesel fuel. Ndayishimiye and Tazerout [10] reported the performance and emissions of crude palm oil (CPO), preheated palm oil and palm oil methyl ester using a single cylinder diesel engine. The results showed, higher brake specific fuel consumption (BSFC), lower brake thermal efficiency (BTE), lower CO and HC and higher NOx compared to the diesel fuel. Liaquat et al. [11] studied PB20 effect during endurance test and claimed lower CO and HC of 11 and 11.71% and higher BSFC and NOx of 3.88 and 3.31% respectively than the diesel fuel. Satyanarayana and Muraleedharan [12] investigated rubber seed oil biodiesel in a single cylinder diesel engine at difference loads and constant speed of 1500 rpm. They reported less torque and brake power (BP), 4.95% lower BTE, higher BSFC, 0.037% lower CO, lower HC and higher NOx compared to diesel fuel. Ramadhas et al. [13] evaluated rubber seed methyl ester in a single cylinder diesel engine for different blends and constant speed of 1500 rpm. The results showed, 12% higher BSFC, lower CO, higher NOx and 17% lower smoke respectively, as compared to diesel fuel. Jena et al. [5] investigated the performance and emissions of mahua-simarouba oil mixture biodiesel using a single cylinder diesel engine. The results showed higher BSFC and NOx, BTE, CO and HC were lower compared to diesel fuel. Haas et al. [14] investigated soybean-soapstock biodiesel in a diesel engine and observed that CO, HC and particle matter (PM) were reduced.

From a literature view, it was clear that there is a major lack of consideration for the production optimization of mixed biodiesel fuel such as rubber seed/palm oil. Therefore in this paper, a new biodiesel fuel with improved properties was developed by blending rubber seed/palm oil at equal blend ratios. The response surface methodology (RSM), in Design Expert 8.0 software was utilized to investigate the effect of different parameters on transesterification process. The experimental design used was Central Composition Design (CCD). Methyl ester at optimized conditions was produced using a two-step (acid esterification and transesterification) process in hydrodynamic cavitation reaction. Thermophysical properties of the produced biofuel and its performance in an IDI diesel engine at full load was investigated.

Table 2.1 Process parameter for transesterification

Process parameters	−2	−1	0	+1	+2
Alcohol to oil molar ratio	4.64	6	8	10	11.36
Catalyst amount (wt%)	0.66	1	1.5	2	2.3
Reaction temperature (°C)	38	45	55	65	72
Reaction time (h)	0.32	1	2	3	3.68

2.2 Rubber Seed/Palm Oil Properties

Rubber seed/palm oil blend acid value of 33.4 mg KOH/g oil was reduced to 1.42 mg KOH/g oil in an acid esterification process using three-neck round bottom flash attached with a condenser to avoid alcohol loss. The temperature was controlled by the help of thermometer. A separating funnel was used to separate the reaction mixture, while a rotatory vacuum evaporator was used to evaporate the methanol and water in the product. As the oil mixture reached the desired temperature, mixture of alcohol (methanol) and catalyst (H_2SO_4) were poured into the flask. Free fatty acid reduction of 1.42 mg KOH/g oil was obtained at alcohol to oil ratio of 15, catalyst loading of 1.5 wt%, reaction temperature of 55 °C and reaction time of 2 h. The oil after the acid esterification process was used for the transesterification process using three-neck round bottom flash of 250 ml. Input factors and their ranges were assigned as, (−1) low level, (+1) high level and on the axial direction were (−2) low level and (+2) high level as presented in Table 2.1. The required temperature, mixing time (mechanical stirrer) and amount of methanol and catalyst (potassium hydroxide) followed the experimental plan as shown in Table 2.2. After the specified time the reaction process was stopped and the product was left for separation gravitationally. Two layers i.e. methyl ester upper and glycerol lower were formed after 12 h. The deionized warm water was used to wash the methyl ester to remove impurities. Finally the produced fatty acid methyl ester (FAME) was stored for properties study.

The preparation of fuels and properties study were carried out at the Universiti Teknologi PETRONAS Centre of Automotive Research and Electric Mobility (CAREM). Three samples were investigated namely a fossil diesel fuel, B5 (5% biodiesel and 95% diesel), B10 (10% biodiesel and 90% diesel) and B20 (20% biodiesel and 80% diesel) vol%. The blends were mechanically stirred for 30 min at 2500 rpm. The properties of the fuels were studied following the ASTM and EN standard methods. Anton Paar (model DMA 4500 M) was used to measure the density following the ASTM D 5002 standard method. Anton Paar, Lovis (model 2000 M/ME) was used to study the viscosity. Cloud point and pour point temperatures were obtained by CPP 5G's analyzer based on ASTM D 2500 and ASTM D 97 test methods respectively. FPP 5G's and 873-CH-9101 Metrohm analyzer were used to measure the cold filter plugging point and oxidation stability following the ASTM D 6371 and EN 14112 standard. The octane number was analyzed using an SHATOX SX-100K) following the ASTM D-613 standard method. The surface ten-

Table 2.2 Transesterification design plan, experimental and predicated FAME yield

xp. run	Alcohol/oil ratio	Catalyst amount (wt%)	Reaction temp. (°C)	Reaction time (h)	Response FAME yield	Predicted FAME yield
1	6	1.00	45	1.00	92	90.88
2	8	2.34	55	2.00	71	72.19
3	6	1.00	65	1.00	97	95.37
4	8	1.50	55	2.00	83	84.17
5	10	2.00	45	1.00	85	84.38
6	8	1.50	55	0.32	72	73.39
7	6	2.00	65	3.00	90	88.37
8	8	1.50	55	3.68	80	81.59
9	10	1.00	65	3.00	70	68.37
10	6	2.00	45	3.00	53	51.88
11	11.36	1.50	55	2.00	88	89.91
12	8	1.50	55	2.00	83	84.17
13	10	2.00	65	1.00	84	82.57
14	4.64	1.50	55	2.00	92	93.99
15	10	1.00	45	3.00	79	78.58
16	8	0.66	55	2.00	70	71.59
17	8	1.50	55	2.00	90	84.17
18	8	1.50	38	2.00	77	77.60
19	8	1.50	55	2.00	84	84.17
20	8	1.50	55	2.00	85	84.17
21	8	1.50	72	2.00	87	89.79

sion was measured using rame hart model 260 following the pendant drop method. Carbon, hydrogen, nitrogen, sulphur and oxygen contents in the biodiesel fuel were determined using CHNS analyser following the classical Pregl-Dumas method. The fatty acid compositions were determined using gas chromatographer (GC) equipped with the Flame Ionization Detector (FID).

2.3 Engine Test Bed

An unmodified multi-cylinder indirect injection diesel engine (IDI), model XLD 418D was used for this investigation. The experimental setup and engine specifications are shown in Fig. 2.1 and Table 2.3. An eddy current dynamometer, model SE 150, water cooled, maximum power of 150 kW, maximum torque of 500 Nm and maximum speed of 8000 rpm was used. The engine and the dynamometer were controlled using engine control unit (ECU) equipped with sensors, logging and a data acquisition device. The experiments started with the engine warm up for about

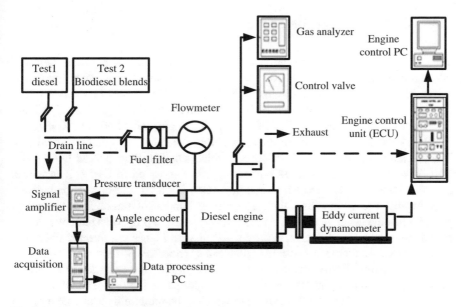

Fig. 2.1 Schematic diagram of the engine testing

Table 2.3 Engine specifications

Engine	Diesel engine
Model	XLD 418D
Type	Four stroke, 44 kW at 4800 rpm
Cylinder number	4 in line OHC, water cooling pressurized circulation
Combustion	IDI, natural aspirated
Bore × stroke	82.5 × 82 mm
Displacement	1753 cc
Compression ratio	21.5:1

30 min using diesel fuel and then the tests were conducted at full (100% load) at speeds of 1000 to 4500 rpm. The engine was flushed with fossil diesel after every fuel change and run for 20 min to insure full exhaustion of the previous sample. The experiments were repeated two times to insure stable reading of torque, engine oil and cooling water inlet and outlet temperatures and fuel mass flow. Emissions such as NOx, CO and CO_2 were measured using a Mur Vario Plus Industrial exhaust gas analyser model 944008 which was calibrated at each fuel changed. The repeated measured data for each blend were averaged prior to their use for the analysis and discussion.

2.4   Biodiesel Properties Analysis

The commercialization of any biodiesel fuel required to meet a set of requirements which are defined in ASTM D 6751 and EN 14214 standard. The fuel properties of mixed biodiesel fuel analysed by using both standards is shown in Table 2.4. The density and acid value of the mixed biodiesel fuel were 0.874 g/m^3 at 15 °C and 0.32 mg KOH/g, whereas for diesel fuel were 0.825 g/m^3 and 0.003 mg KOH/g oil. The acid value and the density of the mixed biodiesel complied with both standards. The viscosity was 4.83 mm^2/s at 40 °C, which satisfied both ASTM D6751 and EN14214 standards. The cetane number which was found to be 52, it is within the both standards. The oxidation stability (OS) is one of the critical parameters affecting the use of biodiesel fuel. The higher content of saturated fatty acid in palm oil has the negative impact on the low temperature properties but better terms of oxidation stability, whereas, rubber seed oil with higher unsaturated fatty acid content makes it suitable for the cold regions but it is not stable to oxidation. Therefore rubber seed/palm oil mixture having ~62% unsaturation has remarkably enhanced the oxidation stability and low temperature properties of the fuel. The OS stability of mixed biodiesel fuel was 8.92 and it is satisfied the minimum requirement of EN 14214 standard. The carbon to hydrogen ratio (C/H) of mixed biodiesel fuel was found to be 6.377 compared to diesel fuel of 6.557.

2.4.1   Effect of Catalyst, Oil to Alcohol Ratio, Reaction Time and Temperature on Fatty Acid Methyl Ester (FAME) Yield

The parametric effects such as reaction temperature, time, catalyst and oil to alcohol ratio on FAME conversion based transesterification are presented in Fig. 2.2a–d. It was observed that by increasing the alcohol to oil ratio and catalyst, the FAME yield decreased as shown in Fig. 2.2a–c. The reason behind is due to the saponification reaction resulting in poor product separation and high glycerol formation [15]. Also, it was noticed that the first 25–30 min of reaction time were enough to achieve the maximum amount of FAME yield, whereas the FAME conversion rate increased as the reaction temperature increased and promoted reaction towards the product side as presented in Fig. 2.2b, d [16]. Ahmad et al. [17] claimed that increasing the alcohol amount increases the ester content to a certain limit before it decreases as the alcohol ratio increases, whereas the higher amount of methanol ratio hinders the glycerol separation, hence lowers the FAME yield.

Table 2.4 Physical properties of methyl ester and blends

Property		Biodiesel	B5	B10	B20	Diesel
Acid value (mg KOH/g oil)		0.32	0.030	0.033	0.063	0.003
Density (kg/m^3) at 25 °C		874	826.3	829.5	834.9	825
Viscosity at 40 °C (mm^2/s)		4.83	3.25	3.38	3.56	3.21
Calorific value (MJ/kg)		38	42.53	42.37	41.93	43.2
Cetane Number		52	47.2	47.86	48.3	47
Oxidation stability (h)		8.92	95.8	92.37	83.26	103.6
Flash Point (°C)		151	75.62	79.86	86.96	72.4
Cloud Point (°C)		5	−15.9	−13.47	−12.15	−17
Pour point (°C)		−1	−30.4	−28.11	−24.23	−32
Clod Filter Plugging Point (°C)		0	−	−	−	−
Surface tension (Nm)		29.3	27.18	27.45	27.63	27.08
Perkin Elmer, CHNS, 2400	Carbon (wt%)	75.38	86.32	85.496	84.272	86.62
	Hydrogen (wt%)	11.82	13.10	13.067	12.524	13.21
	Nitrogen (wt%)	0.07	0.013	0.015	0.021	0.01
	Sulfur (wt%)	0.00	0.156	0.144	0.128	0.16
	Oxygen (wt%)	12.77	2.34	2.5	3.36	0.0

2.4.2 ANOVA (Analysis of Variance) Study for the Transesterification Process

The significance of the transesterification output response was statistically studied using ANOVA test and presented in Table 2.5. The model assumed the output to be significant if P-value is less than 0.05 at 95% confidence level [18]. Data fitting goodness was expressed in terms of the determination coefficient (R^2) and goodness of prediction (adjusted-R^2). The influence of oil molar to alcohol ratio, reaction temperature, time and catalyst ratio on FAME conversion were measured using F-value. The higher the F value of the variable, the higher its influence. In this study B (catalyst) and D (reaction time) were found to be the most influencing variables compared to alcohol to oil ratio (A) and reaction temperature (C) as shown in Fig. 2.3a. it should be noted that the steepest factor is the most influencing compared to others [19].

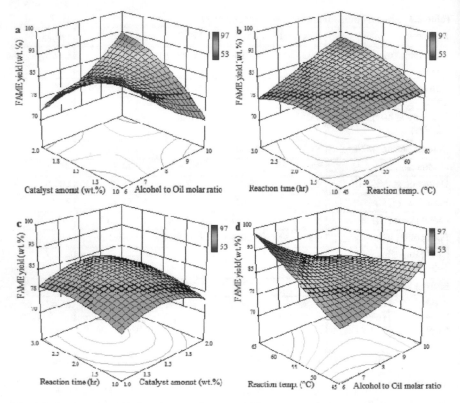

Fig. 2.2 Surface plot of biodiesel conversion, **a** FAME yield versus catalyst amount and alcohol to oil molar ratio, **b** FAME yield versus reaction time and reaction temperature, **c** FAME yield versus reaction time and catalyst amount and **d** FAME yield versus reaction temperature and alcohol to oil molar ratio

The trend between the model predicted and actual values are in good agreement as shown in Fig. 2.3b. The points are close towards the centre linear line. Transesterification optimized condition of 97% conversion yield, were reaction temperature and time of 64 °C and 1 h, catalyst amount of 1.3 and methanol to oil ratio of 6:1. The response equation was produced from the regeration analysis in terms of the actual and the coded terms. The second order coded polynomial equation in terms of most influencing variables on yield is given by Eq. (2.1).

$$\text{FAME yiled} = +88.86 - 6.18 * B + 5.05 * D - 4.61 * B^2$$
$$- 4.93 * D^2 + 9.66 * AB - 5.59 * AC \tag{2.1}$$

Table 2.5 ANOVA based transesterification analysis

Source	Sum of squares	DF	Mean square	F-value	P-value
Model	1764.66	14	126.05	9.24	0.0060
A (alcohol/oil ratio)	0.98	1	0.98	0.072	0.7976
B (catalyst)	216.32	1	216.32	15.86	0.0073
C (temperature)	1.38	1	1.38	0.10	0.7615
D (time)	144.50	1	144.50	10.59	0.0174
A^2	51.71	1	51.71	3.79	0.0994
B^2	317.57	1	317.57	23.28	0.0029
C^2	10.66	1	10.66	0.78	0.4107
D^2	362.92	1	362.92	26.61	0.0021
AB	309.25	1	309.25	22.67	0.0031
AC	250.32	1	250.32	18.35	0.0052
AD	12.80	1	12.80	0.94	0.3701
BC	54.34	1	54.34	3.98	0.0929
BD	0.072	1	0.072	0.01	0.9444
CD	52.79	1	52.79	3.87	0.0967
Residual	81.83	6	13.64		
Lack of fit	80.33	2	40.17	107.11	>0.06
Pure error	1.50	4	0.38		
$R^2 = 0.96$	$R^2_{adj} = 0.85$		Adequate precision = 10.93		

2.5 Engine Performance and Emission Evaluation at Full Load

The variation and averages of engine emissions and performance over the speed range at full load are presented in Fig. 2.4. The brake power (BP) of the tested fuels were ranged from 8.47–39.09, 8.45–37.68, 8.26–36.27 and 7.85–35.79 kW respectively for diesel, B5, B10 and B20. At maximum brake torque speed of 2500 rpm, the BP were 25.30, 24.97, 24.52 and 24.21 kW respectively for diesel, D5, B10 and B20. On average over speed, the BP for diesel, B10 and B20 were 25.78, 25.36 and 25.10 respectively. Compared to diesel fuel, the blends of B5, B10 and B20 produced lower BP of 1.34, 3.1 and 4.3% respectively due to the combined effect of their lower calorific value and higher density and viscosity [20, 21]. Similar results were elucidated by Palash et al. [22] who reported 0.82–3.68% reduction in BP using coconut biodiesel blends. Habibullah et al. [23] reported 3.92–4.71% reduction in BP using coconut and palm biodiesel blends. The BSFC of the biodiesel fueled diesel engine increased as biodiesel percentage increased as explained by Lapuerta et al. [24]. The

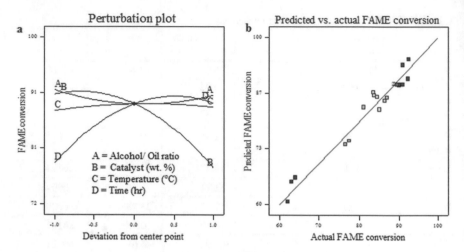

Fig. 2.3 Perturbation plot (**a**), predicted versus actual FAME conversion (**b**)

increase in the BSFC is due to higher density and lower heating value (because of fuel borne oxygen) resulted in more fuel been injected into the combustion chamber compared to neat diesel fuel [25]. The BSFC were in range of 261.42–338.52, 268.03–344.42, 275.25–418.57 and 283.53–473.12 g/kWh. On average the BSFC's were 1.85, 3.76 and 6.39% higher for B5, B10 and B20 respectively compared to diesel fuel.

This finding is concordant with Özener et al. [25] who observed 2–9% increase in BSFC with the use of jatropha methyl ester. The BTE were 1.44, 1.79 and 4.26% lower than diesel fuel. The CO emission of the blends was 11.23–25.84% lower compared to diesel fuel, while CO_2 emission was 0.96–5.01% higher. The reasons were the higher oxygen element and higher cetane number in the biodiesel fuels which enhanced the combustion process [26]. Higher oxygen content ensured complete combustion by allowing more carbon to burn [27], whereas the higher cetane number (higher flame speed and post flame oxidation), prolonged the combustion duration and reduced the possible formation of rich fuel zones and hence reduced CO emission [28]. Fattah et al. [29] also reported a CO reduction of 24.32% with coconut biodiesel blend. NOx, exhaust temperature and O2 of the blends were 1.42–6.22%, 7.94–11.05% and 4.68–13.24 higher compared to diesel fuel respectively. This probably is because of the higher combustion temperature and higher boiling point constituents in the biodiesel fuel which burnt late in the combustion phase [30, 31]. Also, higher carbon double bond increase the free radicals hydrocarbon formation and increase the NO formation [32].

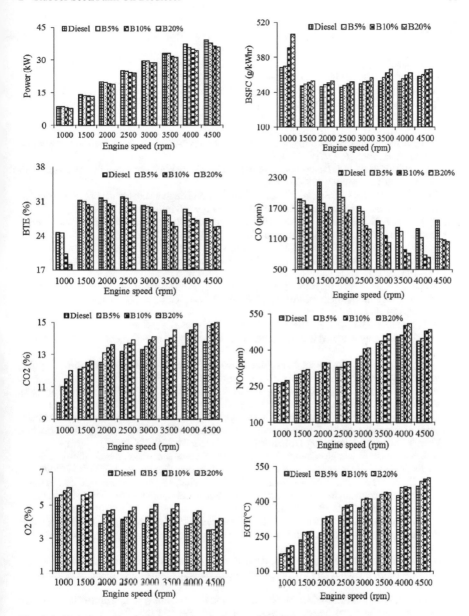

Fig. 2.4 Variation of performance and emissions at full load condition

References

1. N. Wendy, P. Qin, L. Hon Loong, S. Yusup, Supply network synthesis on rubber seed oil utilisation as potential biofuel feedstock. Energy **55**, 82–88 (2013)
2. A.M. Ashraful, H.H. Masjuki, M.A. Kalam, I.M. Rizwanul Fattah, S. Imtenan, S.A. Shahir et al., Production and comparison of fuel properties, engine performance, and emission characteristics of biodiesel from various non-edible vegetable oils: a review. Energy Convers. Manag. **80**, 202–228 (2014)
3. R. Sarin, M. Sharma, S. Sinharay, R.K. Malhotra, Jatropha–Palm biodiesel blends: an optimum mix for Asia. Fuel **86**, 1365–1371 (2007)
4. C.H. Yi, C.T. Han, C.J. Hong, An optimum biodiesel combination: Jatropha and soapnut oil biodiesel blends. Fuel **92**, 377–380 (2012)
5. R.H. Jena, P.C. Jadav, S. Snehal, Performance of a diesel engine with blends of biodiesel (from a mixture of oils) and high-speed diesel. Int. J. Energy Environ. Eng. **4**, 1–1 (2013)
6. H.D. Eka, A.Y. Tajul, N.W.A. Wan, Potential use of Malaysian rubber (*Hevea brasiliensis*) seed as food, feed and biofuel. Int. Food Res. J. **17**, 527–534 (2010)
7. S.N.M. Khazaai, G.P. Maniam, M.H.A. Rahim, M.M. Yusoff, Y. Matsumura, Review on methyl ester production from inedible rubber seed oil under various catalysts. Ind. Crops Prod. **97**, 191–195 (2017)
8. S. Nagaraja, M. Sakthivel, R. Sudhakaran, Combustion and performance analysis of variable compression ratio engine fueled with preheated palm oil—diesel blends. Indian J. Chem. Technol. **20**, 189–194 (2013)
9. O.A. Necati, C. Mustafa, A. Turkcan, C. Sayin, Performance and combustion characteristics of a DI diesel engine fueled with waste palm oil and canola oil methyl esters. Fuel **88**, 629–636 (2009)
10. P. Ndayishimiye, M. Tazerout, Use of palm oil-based biofuel in the internal combustion engines: performance and emissions characteristics. Energy **36**, 1790–1796 (2011)
11. A.M. Liaquat, H.H. Masjuki, M.A. Kalam, M.A. Fazal, A.F. Khan, H. Fayaz et al., Impact of palm biodiesel blend on injector deposit formation. Appl. Energy **111**, 882–893 (2013)
12. M. Satyanarayana, C. Muraleedharan, Investigations on performance and emission characteristics of vegetable oil biodiesels as fuels in a single cylinder direct injection diesel engine. Energy Sour. Part A Recovery Utilization Environ. Eff. **34**, 177–186 (2012)
13. A.S. Ramadhas, C. Muraleedharan, S. Jayaraj, Performance and emission evaluation of a diesel engine fueled with methyl esters of rubber seed oil. Renew. Energy **30**, 1789–1800 (2005)
14. M.J. Haas, K.M. Scott, T.L. Alleman, R.L. McCormick, Engine performance of biodiesel fuel prepared from soybean soapstock: a high quality renewable fuel produced from a waste feedstock. Energy Fuels **15**, 1207–1212 (2001)
15. S. Yusup, A. Bokhari, C. Lai Fatt, J. Ahmad, Pre-blended methyl esters production from crude palm and rubber seed oil via hydrodynamic cavitation reactor. Chem. Eng. Trans. **43**, 517–522 (2015)
16. A. Bokhari, S. Yusup, M.K.R. Nik, J. Ahmad, Blending study of palm oil methyl esters with rubber seed oil methyl esters to improve biodiesel blending properties. Chem. Eng. Trans. **37** (2014)
17. J. Ahmad, S. Yusup, A. Bokhari, R. Nik Mohammad Kamil, Study of fuel properties of rubber seed oil based biodiesel. Energy Convers. Manag. **78**, 266–275 (2014)
18. A. Abuhabaya, J. Fieldhouse, D. Brown, The effects of using biodiesel on CI (compression ignition) engine and optimization of its production by using response surface methodology. Energy **59**, 56–62 (2013)
19. Y. Xingzhong, G.Z. Jia Liu, S. Jingang, T. Jingyi, H. Guohe, Optimization of conversion of waste rapeseed oil with high FFA to biodiesel using response surface methodology. Renew. Energy **33**, 1678–1684 (2008)
20. M.A. Kalam, H.H. Masjuki, M.H. Jayed, A.M. Liaquat, Emission and performance characteristics of an indirect ignition diesel engine fuelled with waste cooking oil. Energy **36**, 397–402 (2011)

21. I.R. Fattah, H. Masjuki, M. Kalam, M. Wakil, A. Ashraful, S.A. Shahir, Experimental investigation of performance and regulated emissions of a diesel engine with *Calophyllum inophyllum* biodiesel blends accompanied by oxidation inhibitors. Energy Convers. Manag. **83**, 232–240 (2014)
22. S. Palash, M. Kalam, H. Masjuki, M. Arbab, B. Masum, A. Sanjid, Impacts of NOx reducing antioxidant additive on performance and emissions of a multi-cylinder diesel engine fueled with Jatropha biodiesel blends. Energy Convers. Manag. **77**, 577–585 (2014)
23. M. Habibullah, H.H. Masjuki, M.A. Kalam, I.M.R. Fattah, A.M. Ashraful, H.M. Mobarak, Biodiesel production and performance evaluation of coconut, palm and their combined blend with diesel in a single-cylinder diesel engine. Energy Convers. Manag. **87**, 250–257 (2014)
24. M. Lapuerta, O. Armas, J. Rodriguez-Fernandez, Effect of biodiesel fuels on diesel engine emissions. Prog. Energy Combust. Sci. **34**, 198–223 (2008)
25. O. Özener, L. Yüksek, A.T. Ergenç, M. Özkan, Effects of soybean biodiesel on a DI diesel engine performance, emission and combustion characteristics. Fuel **115**, 875–883 (2014)
26. A. Sanjid, M.A. Kalam, H.H. Masjuki, M. Varman, N.W.B.M. Zulkifli, M.J. Abedin, Performance and emission of multi-cylinder diesel engine using biodiesel blends obtained from mixed inedible feedstocks. J. Clean. Prod. **112**, 4114–4122 (2016)
27. A.E. Atabani, A.S. Silitonga, I.A. Badruddin, T. Mahlia, H. Masjuki, S. Mekhilef, A comprehensive review on biodiesel as an alternative energy resource and its characteristics. Renew. Sustain. Energy Rev. **16**, 2070–2093 (2012)
28. L.F. Chuah, A.R.A. Aziz, S. Yusup, A. Bokhari, J.J. Klemeš, M.Z. Abdullah, Performance and emission of diesel engine fuelled by waste cooking oil methyl ester derived from palm olein using hydrodynamic cavitation. Clean Technol. Environ. Policy, 1–13 (2015)
29. I.R. Fattah, M.H. Hassan, M.A. Kalam, A.E. Atabani, M.J. Abedin, Synthetic phenolic antioxidants to biodiesel: path toward NO x reduction of an unmodified indirect injection diesel engine. J. Clean. Prod. **79**, 82–90 (2014)
30. K. Muralidharan, D. Vasudevan, Performance, emission and combustion characteristics of a variable compression ratio engine using methyl esters of waste cooking oil and diesel blends. Appl. Energy **88**, 3959–3968 (2011)
31. S.M.R. Ashrafur, H.M. Hassan, M.A. Kalam, M.J. Abedin, A. Sanjid, H. Sajjad, Production of palm and *Calophyllum inophyllum* based biodiesel and investigation of blend performance and exhaust emission in an unmodified diesel engine at high idling conditions. Energy Convers. Manag. **76**, 362–367 (2013)
32. S.K. Hoekman, C. Robbins, Review of the effects of biodiesel on NOx emissions. Fuel Process. Technol. **96**, 237–249 (2012)

Chapter 3
Syngas Dual-Fuelling

Bahaaddein K. M. Mahgoub, Shaharin A. Sulaiman and Suhaimi Hassan

3.1 Introduction

The research for alternative fuel to reduce the dependence on fossil diesel is needed to run internal combustion engines (ICE) with controlled level of exhaust emissions. Therefore many researches were conducted to find suitable renewable fuels with stable prices that can replace fossil fuels [1, 2]. Gaseous fuels are effective alternative sources to replace fossil diesel based on environment and economic consideration, where it emits lower level of nitrogen oxides (NO_X), Sulphur oxides (SO_X) and carbon dioxide (CO_2) [3]. Recently, researchers have paid attention on the use of syngas to run either compression ignition (CI) or spark ignition (SI) engines due to the availability of biomass feedstock for syngas production all over the world. Constituents of syngas that produced through gasification process are varied according to the variation in gasification operating condition and type of biomass that used as feedstock [4]. Table 3.1 lists the constituents of producer gas produced from biomass downdraft gasifier and typical percentage range of each component on a volumetric basis [5, 6].

Using of syngas in ICE to replace diesel is the main subject of this chapter. Syngas is known by its high self-ignition temperature, therefore it is used under dual fuelling mode in CI engines [7]. For syngas dual fuelling in CI engine, two different fuels are

Table 3.1 Constituents of producer gas from biomass gasification [5, 6]

Gas components	CO	H_2	CH_4	CO_2	N_2	Water vapor
Percentage in volume (%)	18–22	15–19	1–5	7–12	45–55	4

B. K. M. Mahgoub (✉) · S. A. Sulaiman · S. Hassan
Universiti Teknologi PETRONAS, 32610 Seri Iskandar, Perak, Malaysia
e-mail: m.bahaa02@uofk.edu

© The Author(s), under exclusive licence to Springer Nature Singapore Pte Ltd., part of Springer Nature 2018
Z. A. Abdul Karim and S. A. Sulaiman (eds.), *Alternative Fuels for Compression Ignition Engines*, SpringerBriefs in Energy, https://doi.org/10.1007/978-981-10-7754-8_3

Fig. 3.1 Conceptual
diagram of dual fuel CI
engine [13]

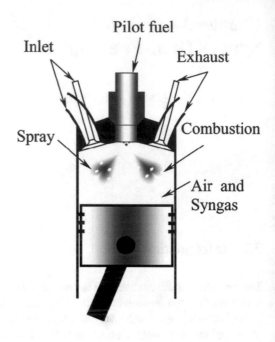

combusted into the combustion chamber namely syngas as primary fuel and diesel
as pilot fuel. The syngas is mixed with air and carbureted at the engine air intake.
The combustion process takes place as in the normal diesel operation, where small
amount of the pilot fuel is needed to combust the air-syngas mixture at the end of
compression stage [8–10]. The advantage of syngas dual fuelling is concluded on
the sustainability of energy and stability of ignition that comes from the use of tow
fuels. In addition, the operation system can be easily switched either for syngas
dual fuelling or normal diesel operation. Furthermore, syngas dual fuelling with
lean mixture could lead to better performance [11]. The syngas dual fuelling in CI
engines has some limitations, where reduction in engine power output is generally
prospected and there is always need for ignition source (diesel). In addition, injection
jet might be overheated for syngas dual fuelling with low diesel flow rate (10–15%)
of the normal flow. Therefore, regular check for the nozzles is needed after 500 h
of syngas dual fuelling operation [12]. Figure 3.1 shows the conceptual diagram of
dual fuelling system for CI engine.

Dual fuel concept was used to run various types of engine with different gaseous
fuels [14–19], but some of them were operating with SI system. Syngas dual fuelling
in SI engines suffers from instability of combustion at high load coditions due to
the syngas fluctuation. Moreover, syngas is a low-energy-density fuel compared to
gasoline therefore a high power degradation is expected for syngas dual fuelling in SI
engines. Syngas dual fuelling in CI engines is more appropriate, but the performance
and emissions of these engines need to be ascertained.

Krishna and Kumar [20] investigated how the syngas dual fuelling effect on a diesel engine operating with syngas produced through coffee husks gasification. Maximum diesel reinstatement of 31% was shown with the application of syngas dual fuelling technique. A gasification gas was used by Sridhar et al. [21] for dual fuelling a CI engine in order to investigate the level of NO_X and CO concentration in the exhaust emissions. Lower levels of NO_X concentration was shown for dual fuelling operation than that for normal diesel operation. On the other hand, higher CO levels were emitted for dual fuelling operation due to ignition inefficiencies.

Sahoo et al. [22] conducted an experimental work to study the performance of syngas dual fuelling in a CI diesel engine with varying H_2:CO ratio and load condition. The H_2:CO ratios were 100:0%, 75:25% and 50:50%. The combustion characteristics, performance and emission of a water cooled, single cylinder DI diesel engine were investigated for the different syngas compositions. Small change on the brake thermal efficiency was presented with varying H_2:CO ratio at low engine loads. However, higher brake thermal efficiency was produced beyond the medium load condition for syngas dual fuelling with higher hydrogen syngas. For syngas dual fuelling with 100% hydrogen syngas at 80% load condition, higher diesel substitution rate, volumetric efficiency and cylinder pressure were presented. For exhaust emission, the level of NO_x concentration was high due to the combustion temperature with the use of 100% H_2. Syngas dual fuelling with 100% hydrogen also emitted lower CO and HC emission compared to the one from other compositions at all load conditions. Increased engine load led to increase the level of CO, HC and NO_x concentration for all examined syngas's.

Earlier studies have explored the potential of syngas as an alternative fuel either for CI or SI engine. They have provided and compared the results based on a variety of syngas compositions at constant engine speed. All these studies were done in order to maximize the engine performance and reduce the level of exhaust emission for dual fuelling operation. It was clear from the reviewed literature that, the main problem of using gasifier-engine systems is the fluctuation of syngas conditions produced from real biomass gasification. So, it would be difficult to assess the effect of producer gas conditions to the performance and emission when combusted in a CI engine under dual fuelling mode. In order to overcome the situation, this research aims at studying of performance and emission of syngas dual fuelling with the use of different compositions of simulated syngases at different engine speeds. Once the problem of syngas composition instability solved, an improvement in the engine performance and emission would be seen by obtaining the most appropriate range of syngas composition and diesel replacement ratio. It is expected that, the presence of hydrogen, carbon monoxide and methane might be the most influential factors affecting the performance and emission for engine operation under syngas dual fuelling with different simulated syngas compositions.

In this chapter, the authors documented a range of simulated syngas compositions under which can get better efficiency and lower emission in syngas dual fuelling for a naturally aspirated CI engine. The objectives of this work in brief are: Firstly to discuss the performance of syngas dual fuelling of a CI engine with different syngas compositions at different diesel replacement ratios and engine speeds; Secondly to

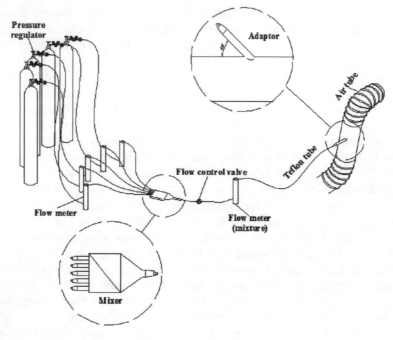

Fig. 3.2 Simulated syngas supply system

analyze the exhaust emission at rated engine performance; Finally to compare the operating ranges of simulated syngas versus pure diesel in a CI diesel engine.

3.2   Simulated Syngas Dual Fuelling Strategy in a CI Engine

In order to deliver the simulated syngas into the engine for dual fueling operation, an experimental rig was connected to the engine air intake without any modification in the engine. The rig consists of five gas cylinders that contains all the syngas constituents which are mixed with air before entering the engine air intake. The syngas constituents flow was controlled by using a flow meter (150 mm) in order to obtain the selected simulated syngas composition. All the gases were mixed on a five way mixer which was fabricated out of carbon steel. The air-simulated syngas mixture supply system is shown in Fig. 3.2.

Table 3.2 Properties of selected compositions of simulated syngas

Composition	N_2	CO_2	CO	H_2	CH_4	LHV (kJ/kg)	ρ (kg/m^3)
A [23]	49	12	25	10	4	4726.19	1.1
B [24]	50.8	9.7	22.1	15.2	1.7	5418.4	0.94
C [25]	38	8	29	19	6	7444.13	0.93

3.3 Selecting of Syngas Composition

As the syngas constituents from real biomass gasification process fluctuates, three different simulated syngas compositions; namely A, B and C were selected to be used for syngas dual fuelling process. The syngas compositions were selected to be representative of biomass gasification process. In order to simulate the real producer gas condition the syngas compositions were selected to be within the typical range of syngas composition. Composition A was selected from Community Power Corporation [23] and composition B was selected from Biomass Energy Foundation (BEF) in the book by Reed and Guar [24], while composition C was selected from previous study by Papagiannakis et al. [25]. The selected simulated syngas compositions are listed in Table 3.2 along with their densities (ρ) and lower heating values (LHV).

3.4 Experimental Set-Up, Procedure and Conditions

The engine test bed along with the instrumentation that was being used in the research is explained in this section. Naturally aspirated, two stroke, single cylinder Tecumseh 5 hp diesel engine was used to investigate the performance and emission of syngas dual fuelling with different types of simulated syngas at different engine speeds and various diesel replacement ratios. Figure 3.3 shows the engine schematic setup, while the technical data of the engine is listed in Table 3.3. The base line test for normal diesel operation was conducted first at engine speed of 2000 rpm. Introducing the simulated syngas into the combustion chamber led to increase in the engine speed. The engine speed was maintained at 2000 rpm and data was recorded. The syngas was varied and further increased while maintaining the engine speed at 2000 rpm in order to investigate the data variation as a function of diesel replacement ratio. The previous procedures were repeated to test the engine at lower speed of 1200 rpm and higher speed of 3000 rpm. The emission measurements were taken for every test at the same time by using multi-component FTIR gas analyzer (GASMET Cr-4000). To ensure the accuracy of the data, routine calibration was mainly done to the dynamometer and the exhaust gas analyzer.

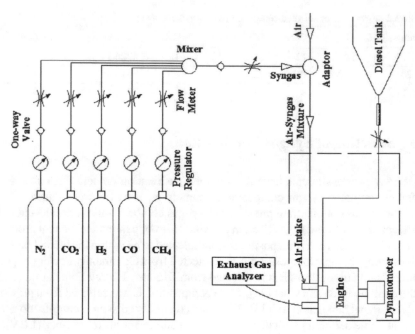

Fig. 3.3 Schematic diagram of the engine test bed

Table 3.3 Engine technical data

Engine type	Single cylinder horizontal P.T.O. shaft
Bore × Stroke (mm)	70 × 60
Piston displacement (cm^3)	230
Compression ratio	17.6:1
Maximum output (HP (kw)/rpm)	4.8 (3.5)/3600
P.T.O shaft rotation	Counter-clockwise facing P.T.O. shaft
Fuel	Diesel light oil
Fuel tank capacity (L)	3.2
Lubricating oil capacity (L)	0.9
Combustion system	Direct injection
Starting system	Recoil starter (electric starter as option)
Dry weight (kg)	29
Dimension L × W × H (mm)	329 × 357 × 402

3.5 Performance of Syngas Dual Fuelling in a CI Engine

This section discuss the results obtained to achieve one of the study objectives which was to determine the effect of syngas composition on the performance characteristics

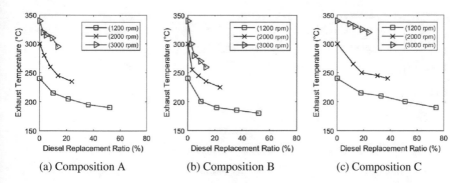

(a) Composition A   (b) Composition B   (c) Composition C

Fig. 3.4 Exhaust temperature versus diesel replacement ratios for syngas dual fuelling

of syngas dual fueling through developing a gaseous fuel supply system in a naturally aspirated CI diesel engine. The engine performance was evaluated in terms of exhaust temperature, engine brake power, brake specific fuel consumption, brake thermal efficiency, fuel air equivalence ratio and volumetric efficiency.

3.5.1 Exhaust Gas Temperature

Figure 3.4 shows the engine exhaust temperature as a function of diesel replacement ratio for syngas dual fuelling with different simulated syngas compositions at various engine speeds. It was observed that increasing the engine speed leads to reduction in diesel replacement ratio, because the diesel flow rate increases for the same flow rate of syngas. The trend was same for all selected simulated syngas compositions. Significant reduction in exhaust temperature was noticed for syngas dual fuelling with composition C due to the ability of engine operation up to high range of diesel substitution.

Reduced exhaust temperature was noticed with the increase of simulated syngas amount for engine dual fuelling at all examined speeds. Syngas dual fuelling with composition C at all examined speeds produced the highest exhaust temperature compared to that of compositions A and B. This is due to the high presence of CO and CH_4 concentration in composition C. The exhaust gas temperature decreases at high engine speed than that at low engine speed because of the high flame temperature caused by the availability of more fuel inside the combustion chamber at the high engine speed.

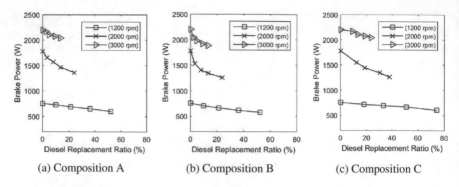

(a) Composition A   (b) Composition B   (c) Composition C

Fig. 3.5 Brake power versus diesel replacement ratios for syngas dual fuelling

3.5.2 Brake Power

Figure 3.5 shows the engine brake power output as a function of diesel replacement ratio for syngas dual fuelling with different simulated syngas compositions at different engine speeds. Higher brake power was produced for engine dual fuelling with all examined simulated syngas compositions at higher engine speed, because more fuel flowing into the combustion chamber at high engine speed.

Syngas dual fuelling with all examined simulated syngas compositions produced lower brake power than normal diesel operation (DR% = 0) due to the lower heating value of syngas compared to diesel. Composition C was the best among all examined simulated syngas compositions in term of produced brake power at dual fuelling operation. This is due to the high presence of CO and CH_4 concentration in composition C. Syngas dual fuelling with composition C also showed higher DR% at 1200 rpm than that for other compositions as shown in Fig. 3.5c. This is due to the lower density of composition C compared to other examined simulated syngas compositions which guarantee the availability of more air for combustion process.

3.5.3 Brake Specific Fuel Consumption (BSFC)

The brake specific fuel consumption as a function of diesel replacement ratio for syngas dual fuelling with different simulated syngas compositions at different engine speeds is shown in Fig. 3.6. Lower value of BSFC was recorded for syngas dual fuelling with all examined simulated syngas composition at 2000 rpm compared to the one for 1200 and 3000 rpm. This is because the heat takes long time to be transferred from the diesel combustion to cylinder walls at low engine speed. While friction losses at high speed decreases the brake torque resulting in higher BSFC.

For engine dual fuelling at 3000 rpm, composition C produced the lowest BSFC compared to other compositions. The low heating value syngas is generally the main

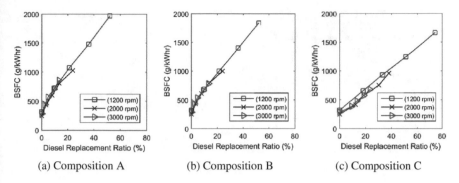

Fig. 3.6 Brake specific fuel consumption versus diesel replacement ratios for syngas dual fuelling

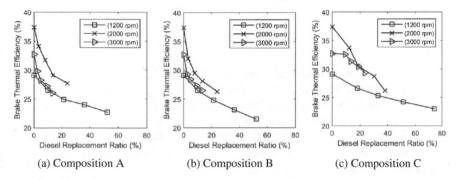

Fig. 3.7 Brake thermal efficiency versus diesel replacement ratios for syngas dual fuelling

reason of higher BSFC with syngas dual fuelling operation compared to that of normal diesel operation.

3.5.4 Brake Thermal Efficiency (BTE)

Figure 3.7 shows the BTE as a function of diesel replacement ratio for syngas dual fuelling with different simulated syngas compositions at various engine speeds. The highest BTE was recorded for engine speed of 2000 rpm for all syngas composi-tions. BTE was lower for syngas dual fuelling beyond 2000 rpm due to the effect of engine friction losses. Higher BTE was produced for richer mixture because the mixture properties approach to those of ideal gases. Therefore, syngas dual fuelling at 2000 rpm showed the highest values of BTE as shown in Fig. 3.7. Syngas dual fuelling generally leads to low thermal efficiency because of the presence of CO_2 in syngas composition [26]. For syngas dual fuelling at higher engine speed, the high of presence of CO and CH_4 content in composition C resulted in slow reduction in the BTE compared to other compositions.

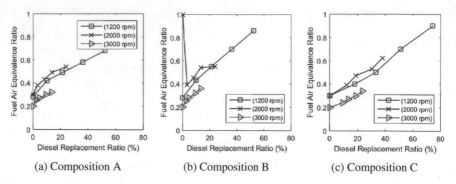

Fig. 3.8 Fuel air equivalence ratio versus diesel replacement ratios for syngas dual fuelling

3.5.5 Fuel Air Equivalence Ratio

Figure 3.8 depicts the fuel air equivalence as a function of diesel replacement ratio for syngas dual fuelling with different simulated syngas compositions at various engine speeds. Increasing the engine speed leads to increased fuel consumption which result in higher fuel air equivalence ratio. Fuel/Air equivalence ratio also increased with the increase of syngas concentration inside the combustion chamber increased for syngas dual fuelling at all examined engine speeds. This is because the air was replaced by syngas which resulted in higher actual fuel/air ratio.

The highest fuel air equivalence ratio was shown for syngas dual fuelling for all examined simulated syngas compositions at 2000 rpm where the engine produce its maximum torque. For syngas dual fuelling at 3000 rpm, the higher air flow resulted in a lower fuel air equivalence ratio accompanied by leaner combustion.

3.5.6 Volumetric Efficiency

The volumetric efficiency as a function of diesel replacement ratio for syngas dual fuelling with different simulated syngas compositions at various engine speeds is shown in Fig. 3.9. It is shown that, higher volumetric efficiency was resulted at high engine speed because more air has been sucked into the combustion chamber due to the high vacuum at engine air intake. Syngas dual fuelling resulted in higher volumetric efficiency than that of normal diesel operation. This is due to the lowering in intake air temperature caused by the low heat transfer rate from engine parts to the air. Furthermore, the volumetric efficiency was higher for syngas dual fuelling operation because air-syngas mixture was inducted into the combustion chamber instead of the air alone as for normal diesel operation. Syngas dual fuelling with composition B showed the highest volumetric efficiency because of the lower heating for composition B compared to other compositions. The low density of Composition C has

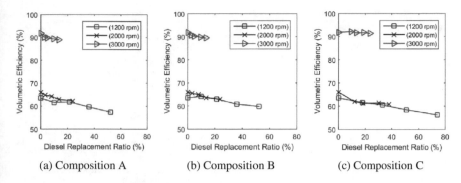

Fig. 3.9 Volumetric efficiency versus diesel replacement ratios for syngas dual fuelling

resulted in the highest volumetric efficiency for syngas dual fuelling at 3000 rpm, where more air was available as interpreted from Fig. 3.9.

It can be concluded from the performance results that, lower power output, exhaust gas temperature and brake thermal efficiency was obtained for syngas dual fuelling compared to normal diesel operation. While the brake specific fuel consumption and volumetric efficiency were higher for syngas dual fuelling. The presence of syngas constituents have affected the engine performance, especially H_2, CO and CH_4 because they are the combustible components in syngas composition. Syngas dual fuelling with composition C showed the maximum diesel replacement (74.2%) at 1200 rpm. The presence of noncombustible gases in syngas, namely CO_2 and N_2 significantly affect the density of syngas, which has an impact on the inhaled air for combustion and emission level.

3.6 Exhaust Emission from a CI Engine Under Syngas Dual Fuelling

The engine exhaust emission was evaluated in terms of CO_2, CO, UHC and NO_X. Generally, high level of CO_2 emission is an indicator for good combustion. While the NO_X concentration is the result of combustion temperatures rise or misfire. When the fuel incompletely combust, higher levels of CO and UHC are presented.

3.6.1 Carbon Dioxide

CO_2 concentration as a function of diesel replacement ratio for syngas dual fuelling with different simulated syngas compositions at various engine speeds is given in Fig. 3.10. Increasing the engine speed leads to higher cylinder temperature which facilitates the oxidation process and formation CO_2 emission. The highest level of

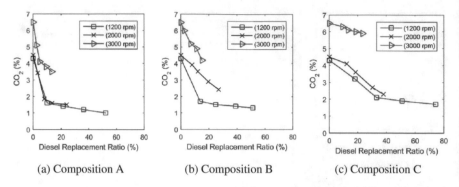

(a) Composition A   (b) Composition B   (c) Composition C

Fig. 3.10 CO_2 concentration versus diesel replacement ratios for syngas dual fuelling

CO_2 concentration was emitted for syngas dual fuelling at 3000 rpm for all examined simulated syngas compositions.

It was observed that, increasing the syngas concentration into the combustion chamber results in reduction on the level of CO_2 emission. This is because some amount of air was displaced by simulated syngas when introduced through the engine air intake. Lower level of CO_2 emission was recorded for composition A at all engine speeds due to its higher density which led to big air displacement.

3.6.2 Carbon Monoxide

Figure 3.11 provides the variation of CO concentration as a function of diesel replacement ratio for syngas dual fuelling with different simulated syngas compositions at various engine speeds. It was observed that, the level of CO concentration increases as the engine speed increases for syngas dual fuelling. This is because of the time shortens for oxidization of CO to CO_2 at high engine speeds. Another reason for higher CO emission with syngas dual fuelling is the lower heating value of syngas that led to incomplete combustion. Therefore, the highest level of CO concentration was shown for syngas dual fuelling with composition A owing to its higher presence of CO.

3.6.3 Unburnt Hydrocarbons

Figure 3.12 shows the concentration of UHC as a function of diesel replacement ratio for syngas dual fuelling with different simulated syngas compositions at various engine speeds. Syngas dual fuelling emitted the highest level of UHC at 3000 rpm for all examined simulated syngas compositions. This is because of the lack of oxygen

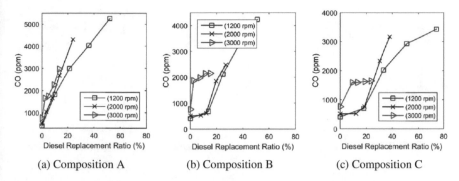

Fig. 3.11 CO concentration versus diesel replacement ratios for syngas dual fuelling

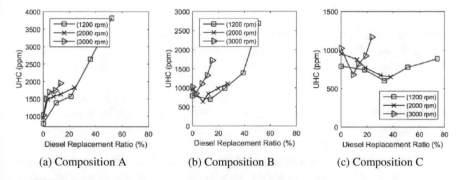

Fig. 3.12 UHC concentration versus diesel replacement ratios for syngas dual fuelling

availability to complete the combustion that might be consumed into the CO and CO_2 formation as interpreted from Fig. 3.10.

Syngas dual fuelling with composition A showed the highest UHC due to the reduction in oxygen available for combustion that cause by higher density of composition A. It was noticed that, rigorous combustion has occurred at the early stage of combustion for syngas dual fuelling with compositions B and C which resulted in lower level of UHC concentration. Then the combustion rate has decreased and the level of UHC concentration gradually increased to a higher level.

3.6.4 Oxides of Nitrogen

NO_X emission as a function of diesel replacement ratio for syngas dual fuelling with different simulated syngas compositions at various engine speeds is shown in Fig. 3.13 for each syngas composition at different engine speeds. It was clearly seen that, higher engine speed results in higher level of NO_X concentration. The main reason for NO_x formation at high engine speed is the higher combustion temperature

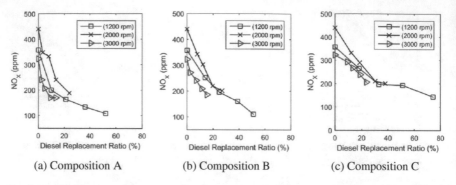

(a) Composition A   (b) Composition B   (c) Composition C

Fig. 3.13 NO_X concentration versus diesel replacement ratios for syngas dual fuelling

and availability of oxygen into the combustion chamber. Nevertheless, syngas dual fuelling showed the lowest level of NO_x concentration at higher speed of 3000 rpm. This is because all the oxygen was consumed in the CO formation as shown in Fig. 3.11.

Significant reduction in the level of NO_X concentration was noticed for syngas dual fuelling due to the dilution effect for nitrogen in the air by syngas mixing. The reduction of oxygen concentration is another reason for the decline of NO_X concentration level for syngas dual fuelling operation. Hence, in this study, the lowest level of NO_X concentration was recorded for syngas dual fuelling with composition A due to its lower heating value that leads to lower the combustion temperature. The highest level of NO_X concentration was shown by composition C due to its high hydrogen content that led to high combustion temperature as indicated by the measured exhaust temperatures.

It can be concluded from the emissions results that, hydrogen presence in syngas composition plays very important role in controlling the level of CO and UHC concentration. Therefore, syngas dual fuelling with Composition C emitted the lowest level of CO and UHC. In addition, it produced the highest brake power output with the lowest brake specific fuel consumption with the maximum diesel replacement ratio.

3.7 Imitated Syngas Operations Range Versus Pure Diesel Fuel

The test results on performance and emission for syngas are compared with the results for diesel fuel only. Table 3.4 shows the comparison between the operating ranges of imitated syngas versus pure diesel fuel.

Table 3.4 Comparison of imitated syngas operations versus pure diesel fuel

Syngas composition	Engine speed (rpm)	Operating range
Composition A	1200	The lowest level of CO_2 concentration was emitted for syngas dual fuelling with composition A at DR of 52.2%
	2000	The lowest level of CO_2 concentration was emitted for syngas dual fuelling with composition A at DR of 23.9
	3000	The lowest exhaust temperature along with the highest volumetric efficiency was shown for syngas dual fuelling with composition A at DR of 13.3%
Composition B	1200	The lowest level of CO_2 concentration was emitted for syngas dual fuelling with composition B at DR of 51.1%
	2000	The lowest exhaust temperature along with the highest volumetric efficiency was shown for syngas dual fuelling with composition B at DR of 26.9%
	3000	The lowest exhaust temperature was shown for syngas dual fuelling with composition B at DR of 15.2%
Composition C	1200	The highest level of power output, and BTE with lowest BSFC were presented for syngas dual fuelling with composition C at DR of 74.2%
	2000	The highest level of power output, and BTE with lowest BSFC and UHC were presented for syngas dual fuelling with composition C at DR of 38%
	3000	The highest level of power output, volumetric efficiency and BTE with lowest BSFC and UHC were presented for syngas dual fueling with composition C at DR of 23.5%

References

1. N. Kapilan, T.P. Ashok Babu, R.P. Reddy, Improvement of performance of dual fuel engine operated at part load. Int. J Automot. Mech. Eng. (IJAME) **2**, 200–210 (2010)
2. M.A. Rahman, A. Ruhul, M. Aziz, R. Ahmed, Experimental exploration of hydrogen enrichment in a dual fuel CI engine with exhaust gas recirculation. Int. J. Hydrogen Energy **42**, 5400–5409 (2017)
3. A. Henham, M. Makkar, Combustion of simulated biogas in a dual-fuel diesel engine. Energy Convers. Manag. **39**, 2001–2009 (1998)
4. K. Tomishige, M. Asadullah, K. Kunimori, Syngas production by biomass gasification using $Rh/CeO_2/SiO_2$ catalysts and fluidized bed reactor. Catal. Today **89**, 389–403 (2004)
5. N.Z. Shilling, D.T. Lee, IGCC-clean power generation alternative for solid fuels. *PowerGen Asia, Ho Chi Minh City, Vietnam, September* (2003), pp. 23–25

6. A. Mohod, S. Gadge, V. Mandasure, Liberation of carbon monoxide through gasifier IC engine system. IE (I) J. ID **84**, 27–29 (2003)

7. B.M. Thayagarajan, A combustion model for a dual fuel direct injection diesel engine, in *The 1st International Symposium on Diagnostics and Modeling of Combustion in Internal Combustion Engines (COMODIA)*, Tokyo, Japan (1985) pp. 607–614

8. A.S. Bika, L.M. Franklin, D.B. Kittelson, in *Hydrogen as a Combustion Modifier of Ethanol in Compression Ignition Engines*. SAE Technical Paper 0148-7191 (2009)

9. S. Lambe, H. Watson, Low polluting, energy efficient CI hydrogen engine. Int. J. Hydrogen Energy **17**, 513–525 (1992)

10. N. Saravanan, G. Nagarajan, An insight on hydrogen fuel injection techniques with SCR system for NOx reduction in a hydrogen–diesel dual fuel engine. Int. J. Hydrogen Energy **34**, 9019–9032 (2009)

11. G.A. Karim, Combustion in gas fueled compression: ignition engines of the dual fuel type. Trans.-Am. Soc. Mech. Eng. J. Eng. Gas Turbines Power **125**, 827–836 (2003)

12. K. Von Mitzlaff, *Engines for Biogas: Theory, Modification, Economic Operation*, (Federal Republic of Germany: Friedr. Vieweg & Sohn Verlagsgesellschaft mbH, 1988)

13. C. Garnier, A. Bilcan, O. Le Corre, C. Rahmouni, in *Characterisation of a Syngas-Diesel Fuelled CI Engine*. SAE Technical Paper 0148-7191 (2005)

14. Y. Ando, K. Yoshikawa, M. Beck, H. Endo, Research and development of a low-BTU gas-driven engine for waste gasification and power generation. Energy **30**, 2206–2218 (2005)

15. Y. Yamasaki, G. Tomatsu, Y. Nagata, S. Kaneko, Development of a small size gas engine system with biomass gas (combustion characteristics of the wood chip pyrolysis gas). Combust. **2013**, 08–24 (2007)

16. F.Y. Hagos, A.R.A. Aziz, S.A. Sulaiman, Trends of syngas as a fuel in internal combustion engines. Adv. Mech. Eng. **2014** (2014)

17. D. Barik, S. Murugan, Investigation on combustion performance and emission characteristics of a DI (direct injection) diesel engine fueled with biogas–diesel in dual fuel mode. Energy **72**, 760–771 (2014)

18. S. Di Iorio, A. Magno, E. Mancaruso, B.M. Vaglieco, Analysis of the effects of diesel/methane dual fuel combustion on nitrogen oxides and particle formation through optical investigation in a real engine. Fuel Process. Technol. **159**, 200–210 (2017)

19. S.S. Kalsi, K. Subramanian, Effect of simulated biogas on performance, combustion and emissions characteristics of a bio-diesel fueled diesel engine. Renew. Energy **106**, 78–90 (2017)

20. K.S. Krishna, K.A. Kumar, A study for the utilization of coffee husk in diesel engine by gasification, in *Proceedings of Biomass Gasification Technology*, India, 55–58 (1994)

21. G. Sridhar, S. Dasappa, H.V. Sridhar, P.J. Paul, N.K.S. Rajan, in *Gaseous Emissions Using Producer Gas as Fuel in Reciprocating Engines*. SAE Technical Paper. https://doi.org/10.4271/2005-01-1732 (2005)

22. B.B. Sahoo, N. Sahoo, U.K. Saha, Effect of H_2:CO ratio in syngas on the performance of a dual fuel diesel engine operation. Appl. Therm. Eng. **49**, 139–146 (2011)

23. C.P. Corporation products: BioMax [Online], Available: http://www.gocpc.com (2009)

24. T.B. Reed, S. Gaur, *A survey of biomass gasification 2000: gasifier projects and manufacturers around the world* (Biomass Energy Found., USA, 1999)

25. R.G. Papagiannakis, C.D. Rakopoulos, D.T. Hountalas, E.G. Giakoumis, Study of the performance and exhaust emissions of a spark-ignited engine operating on syngas fuel. Int. J. Altern. Propul. **1**, 190–215 (2007)

26. P.M. Duc, K. Wattanavichien, Study on biogas premixed charge diesel dual fuelled engine. Energy Convers. Manag. **48**, 2286–2308 (2007)

Chapter 4
Water-in-Diesel Emulsions—Fuel Characteristics

Z. A. Abdul Karim and Mohammed Yahaya Khan

4.1 Introduction

Emission regulations continue to become more stringent, exacerbating the need to find a solution for the emission of gaseous pollutant and particulate matter from diesel engines. Despite the fact that diesel engines offer higher efficiency and fuel economy, they emit pollutants, i.e. particulate matter (PM), nitrogen oxides (NOx), sulphur oxides (SOx), carbon monoxides (CO) and carbon dioxides (CO_2). Several techniques have been used to improve engine performance as well as reducing the NOx and soot emission which requires retrofitting of the existing engine. Adding water to the fuel is a fuel based solution and can be done through several means including injecting water into the combustion chamber using an isolated injector, spraying water into the intake air and water with diesel emulsion [1]. Introduction of water into an engine has been proven to reduce emission through cooling effect of the in-cylinder peak temperature [2, 3]. However, the presence of water in the fuel blend decreases the engine power output, which is generally inversely proportional to the water concentration of the emulsion, due to the lowering of the fuel's heating value compared to neat diesel fuel [4]. In addition to the tendency of the water and fuel to be separated, the presence of water leads to engine parts corrosion when there is direct contact of water with the part surfaces.

On the other hand, water embedded in the diesel fuel can also be used to break the bigger fuel particle into a smaller size due to the difference in evaporation rate of these liquids which leads to a complete combustion of the fuel in the engine, this phenomena is known as microexplosion [5]. Also the presence of water during the

Z. A. Abdul Karim (✉) · M. Y. Khan
Universiti Teknologi PETRONAS, 32610 Seri Iskandar, Perak, Malaysia
e-mail: ambri@utp.edu.my

M. Y. Khan
e-mail: mohammedyahayakhan@yahoo.com

© The Author(s), under exclusive licence to Springer Nature Singapore Pte Ltd., part of Springer Nature 2018
Z. A. Abdul Karim and S. A. Sulaiman (eds.), *Alternative Fuels for Compression Ignition Engines*, SpringerBriefs in Energy, https://doi.org/10.1007/978-981-10-7754-8_4

intensive combustion period seems to diminish the rate of formation of soot particles, which could be attributed to the microexplosion phenomenon [6, 7]. It was found that physics and chemical kinetics of the combustion are influenced by the presence of water vapor in fuel. Water vapor reduces the combustion temperature and changes the chemical composition of reactants to control the NOx and PM formations on the combustion inside the cylinder [4]. Consequently, the exhaust emissions from the engine will reduced according to the percentage amount of water in the diesel. Hence, Water-in-Diesel Emulsion (WiDE) fuel is viewed as an alternative fuel to improve engine efficiency and reduce exhaust emissions, particularly nitrogen oxides (NOx) and particulate matter (PM) [8, 9]. Not only does this alternative fuel reduces exhaust emissions but it also saves fuel [10]. Various studies reported the success of WiDE in solving the emission issue through the microexplosion phenomenon [11, 12] however, the extent of the efficacy of emulsified fuel on the engine performance is not precisely known. This is due to the inadequate understanding on how WiDE behaves under combustible condition as well as the complexity of engine operating parameters. By understanding the conditions of the fuel that would affect these processes will result in a fuel that would improve the degree of efficiency of the injection and combustion mechanism.

4.2   Water-in-Diesel Emulsion

An emulsion consists of two incomplete immiscible liquids (oil and water), with one of the liquids formed as small spherical droplets or dispersed phase distributed in the other liquid present in continuous phase [13]. These type of emulsions are called as two phase emulsions and there are two basic forms of two phase emulsion. The first is the oil-in-water (O/W) emulsion in which oil droplets are dispersed and encapsulated within the water column. The second is the water-in-oil (W/O) emulsion in which droplets of water are dispersed and encapsulated within the oil. Figure 4.1 shows the concept of two-phase oil-in-water emulsions and water-in-oil [5]. For either type of stable emulsion to form, three basic conditions must be met [14], such as (i) the two liquids must be immiscible or mutually insoluble in each other, (ii) sufficient agitation must be applied to disperse one liquid into the other and (iii) an emulsifying agent (emulsifier) or a combination of emulsifiers must be present. Emulsifier is a compound that reduces the surface tensions between two immiscible fluids. They contain both hydrophobic (the tails) and hydrophilic groups (the heads). For a best formation of emulsions, hydrophilic (water liking)—lipophilic (oil liking) balance (HLB) score is developed. Low HLB tends to make water-in-oil-emulsion while those with a high HLB are more hydrophilic and tend to make oil-in-water-emulsion. The value of HLB ranges from 1 to 20.

Emulsifiers used for the formation of water-in-diesel emulsion fuel should burn easily with no soot and should be free of sulphur and nitrogen [15]. Furthermore, they should have no impact on the physicochemical properties of the fuel. Emulsifiers from the aliphatic hydrocarbon family are the best candidates to be used as

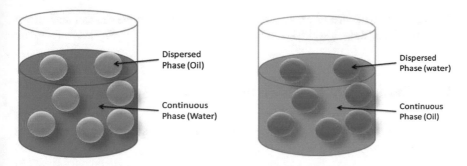

Fig. 4.1 Concept of two-phase oil-in-water and water-in-oil emulsions [5]

emulsifiers. Usually the amount of emulsifiers used for emulsification is in the range of 0.5–5% by volume. According to Huo et al. [16] and Wang et al. [17] in their separate studies they concluded that a blend of two emulsifier gave effective results in producing stable emulsions compared to a single emulsifier. Some commercially available emulsifiers can be easily found in the websites with their HLB values. The stability of the diesel emulsion is affected mainly by the emulsification technique, emulsification duration, volume fraction of water (dispersed phase), viscosity of continuous phase (diesel oil), stirring speed (or ultrasonic frequency) and concentration of emulsifiers. The experimental work by Chen and Tao [18] studied the effect of emulsifier dosage, oil-water ratio, stirring speed and emulsifying temperature on the stability of water-in-diesel emulsion produced using mechanical agitator. They reported that an increase in oil to water ratio, stirring speed and duration had positive influence on stability, whereas an increase in emulsifying temperature showed negative impact. Better engine performance with less CO emissions was reported by Lin and Chen [19] with the application of ultrasonic vibrator to produce the emulsion compared to the emulsion prepared by mechanical agitation. In a different study by Lin and Wang [20] on the effect of speed of mechanical homogenizer machine, and it was demonstrated that the speed had a visible impact on the diameter of liquid droplets. The result showed an increase in the stirring speed resulted in smaller liquid droplets. In addition, the selections of suitable emulsifiers, the choice of a suitable agitator frequency and agitation time have also been identified as equally important parameters in the formation of stable emulsified fuels [21].

A good fuel for compression ignition engines should acquire all the positive characteristics of diesel fuel i.e. short ignition timing, high cetane number, a suitable fuel viscosity and volatility [22]. These physiochemical characteristics greatly affect the fuel injection and combustion process in the engine. For example, the density of fuel would affect the mixing process in the combustion chamber and the viscosity would then affect the injection system. Another important characteristics of an emulsion is its droplets size distribution. Stable emulsion would have a well-distributed size of droplets. The preferable diameters of the small water droplets are in the range of between 0.05–1 μm [4]. A proper HLB value of the emulsifiers is a key factor where the molecular structure of emulsifiers had a great effect on the droplet size of the final emulsions [23].

4.3   Production of Emulsion by Mechanical Agitator

A case study on preparing water-in-diesel emulsion using mechanical agitator is explained below:

i.   The emulsifier is prepared for two different HLB values of 5 and 6.3. A blend of two commercial emulsifiers such as Tween 85 with an HLB value of 11 and Span 80 with an HLB value of 4.3 were used as stabilizer. Span 80 (sorbitan monoleate) and Tween 85 (polyoxyethylene sorbitan trioleate) were used as received in preparing the emulsions. These emulsifiers are lipophilic and hydrophilic type of emulsifiers which increases the diesel/water affinity and help to reduce the interfacial tension and hence increases the emulsion stability. In order to avoid any contamination or impurities in the emulsions, distilled water is always preferred in emulsion blending.

ii. The required volume of distilled water is measured and transferred into the mixing container and then, the emulsifier of HLB value of 5 is measured and transferred using tranferpette into the container that are already filled with water.

The required HLB values were obtained by combining known amount of Span 80 and Tween 85. Equation 4.1 is used to determine the amount of Span 80 to be mixed with Tween 85 in order to obtain the desired HLB value of the emulsifier.

$$\%A = 100 * (x - HLB_B)/(HLB_A - HLB_B) \tag{4.1}$$

where,

HLB_A HLB of the emulsifier A
HLB_B HLB of the emulsifier B
x Desired HLB value
$\%A$ Amount of emulsifier A required
$\%B$ ($100\%-\%A$) Amount of emulsifier B required

iii. Diesel is last measured and transferred into the container having the mixture of water and emulsifier. The mixture is mixed at a fixed at 1500 rpm for 15 min at room temperature.

Table 4.1 shows four water-in-diesel emulsions having different, different water content of 9, 12, 15 and 18% while maintaining the HLB value of 6.31 and an emulsifier dosages of 10%. These H_2O percentages were selected based on the French standard which minimum percentage of water is 9% and a maximum of 15%.

Table 4.1 Water-in-diesel emulsion (WiDE) preparation matrix for a HLB value of 6.31

% of water	For 4 L of WiDE fuel			
	Diesel (in mL)	Water (in mL)	Emulsifier (mL)	
			Span 80	Tween 85
9	3604	360	25.2	10.8
12	3472	480	33.6	14.4
15	3340	600	42.0	18.0
18	3208	720	50.4	21.6

Table 4.2 Chemical characteristics and properties of emulsified fuel and commercial diesel fuel

Fuel	Water content (%)	Calorific value (J/kg)	Carbon (%)	Hydrogen (%)	Nitrogen (%)	Sulfur (%)
Diesel	–	45,135	86.10	12.92	0.06	–
Emulsion-1	9	38,989	76.3	12.493	1.0	0.064
Emulsion-2	12	37,904	73.39	12.631	1.1	0.062
Emulsion-3	15	36,432	69.14	12.265	0.910	0.718

4.3.1 Water-in-Diesel Emulsion Properties

Physical characterization of WiDE samples such as distributed water droplets size and distribution, density at 15 °C, and viscosity at 40 °C are characteristics that influence the injection, mixing and combustion process of these fuel blends in diesel engine. In addition to the above properties, surface tension was also measured as this property influences the stability of the emulsion. Table 4.2 depicts the comparison of elemental chemical properties between the water-in-diesel emulsion and commercial diesel. The properties of the fuels for density, kinematic viscosity and surface tension for diesel and WiDE shown in Table 4.1 are presented in Table 4.3. It can be observed from Table 4.3 that the density and kinematic viscosity of WiDE are comparatively higher than diesel and the increase in density of the emulsions was found to be directly proportionate to increase in water content. It is due to the presence of higher density water in the blend and increased in the number of the disperse phase. In case of surface tension of the emulsions, the presence of water had insignificant influence. The calorific value of WiDE fuels is however, decreased as the percentage of water content increased. This can be attributed that water has no calorific value and increased in water content reduced the overall calorific values of the emulsions.

For the different emulsifier HLB values, the effect of increase in percentage of water on the density and viscosity of the emulsions are shown in Figs. 4.2 and 4.3. It was found that the density the viscosity of the emulsions are found to be increasing with increase in water content. Also, the HLB of the emulsifier has significant influence on the emulsions.

Table 4.3 Properties water-in-diesel emulsion (WiDE)

% H_2O	Density@15 °C (kg/m^3)	Kinematic viscosity (mm^2/s)	Surface tension @ 20 °C (mN/m)
HLB value of 5, 15% emulsifier from H_2O			
9	865.36	5.75	24.34
12	871.84	7.01	24.86
15	874.23	7.37	25.66
18	881.97	9.5	25.03
HLB value of 6.31, 15% emulsifier from			
9	868.37	6.35	24.59
12	875.9	7.86	24.58
15	879.27	9.2	24.84
18	8856	13.25	25.54

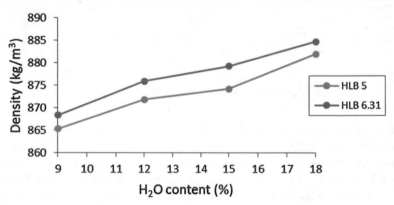

Fig. 4.2 Water content percentage against density (kg/m^3) graph

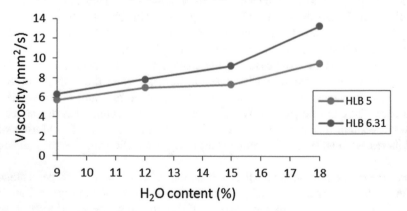

Fig. 4.3 Water content percentage against viscosity (mm^2/s) graph

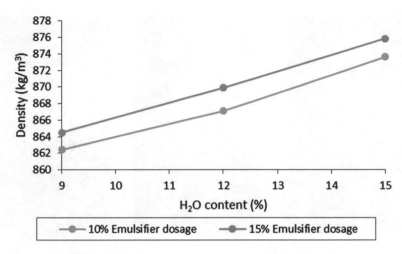

Fig. 4.4 Emulsifier dosage percentage against density (kg/m^3) graph

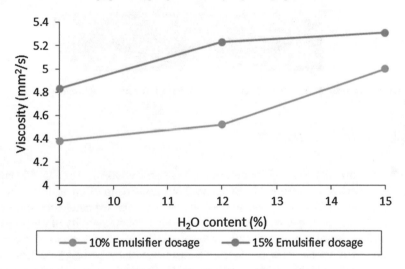

Fig. 4.5 Emulsifier dosage percentage against viscosity (mm^2/s) graph

Figures 4.4 and 4.5 show the effect of increasing emulsifier dosage (i.e., 10 and 15%) on emulsion's density and viscosity. It is clear from the figure that the density of the water-in-diesel emulsion was slightly increased with an increase in emulsifier dosage. Similarly, the influence of emulsifier dosage on emulsion viscosity is also observed since the increase in emulsifier volume resulted in increased viscosity of the emulsions.

Fig. 4.6 Comparison of **a** unstable, **b** unstable with water layer and **c** stable emulsion

4.3.2 Emulsion Stability

Figure 4.6 shows the comparison between (a) unstable emulsion, (b) unstable emulsion with water layer and (c) stable emulsion. The standard procedure to determine the stability of the emulsion is by visual observation. The blended emulsions were kept motionless at room temperature in clear glass bottles and the stability of the prepared water-in-diesel emulsion was visually observed for a period of 2 weeks–3 months.

As soon as the emulsion is mixed it starts to change due to several time dependent processes, among which creaming, sedimentation, Ostwald ripening flocculation and coalescence appear to be most important [24, 25]. Creaming and sedimentation is otherwise called as de-emulsification processes which is the separation of emulsion into two emulsions. The process is an outcome of external forces usually gravitational. When such forces exceeds the thermal forces of the droplets (Brownian motion) a concentration gradient builds up in the mixture which cause the larger droplets moves either to the top (if the dispersed phase is less dense than the oil) or to the bottom (if the dispersed medium, in this case water is more dense than the continuous the diesel).

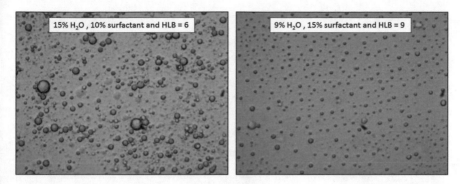

Fig. 4.7 Sample water particle distribution for WiDE

4.3.3   *Water Droplets Diameter and Distribution*

The dispersed water particles size and distribution were examined using a digital microscope at 500× magnification, and the images were captured using a digital camera. Software associated with the digital microscope (MOTIC Image Plus 2.0) was used to post processing of the emulsion images for determining the sizes of the distributed water droplets. The size and the distribution of the water droplets in the emulsions were calculated by using Eq. 4.2,

$$D_{32} = \sum_i \left(n_i \times D_i^3 \right) / \sum_i \left(n_i \times D_i^2 \right) \tag{4.2}$$

where D_i is the diameter of the droplet, and n_i is the total number of droplets having the same diameter.

Figure 4.7 shows the water particle distribution for WiDE with 15% water content stabilized by 10% emulsifier dosage with an HLB value of 6 and 9% water content stabilized by 15% emulsifier dosage with an HLB value of 9. For the two selected emulsifier dosages (10 and 15%), higher emulsifier dosage caused finer water particle size and homogenized a wide range of Sauter Mean Diameter (SMD) was seen for WiDE with 9, 12 and 15% water contents as shown in Fig. 4.8. It shows that the same emulsifier dosage with increase in the water content percentage, (i.e.,) for emulsion with 10 and 15% emulsifier dosage, the difference in the SMD was found to be minimum at 12% water content and as for 9% water content and 15% water content show high differences in SMD value. For the 10% emulsifier dosage, the SMD was found to increase as the water content increases. As for 15% emulsifier dosage, it shows inconsistency in SMD value when the water content increases. Hence, for the emulsions examined it can be concluded that the SMD are slightly influenced by the emulsifier dosage.

With reference to Fig. 4.9, it appears that the optimum reductions in SMD were achieved with 15% water content when the emulsifier dosage increases. As for WiDE with water content of 12%, when the emulsifier dosage increases the SMD value was

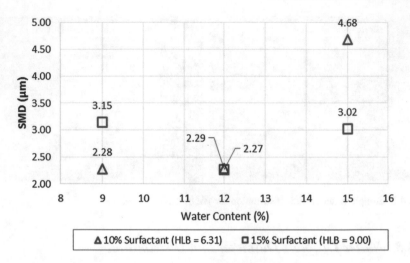

Fig. 4.8 Sauter mean diameter (μm) versus emulsions water content (%)

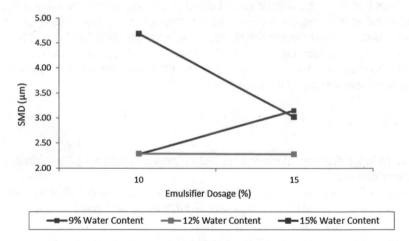

Fig. 4.9 Sauter mean diameter (μm) against emulsions emulsifier dosage (%)

at very minimal reductions or almost consistent, and lastly for water content of 9%, it shows an increment in SMD value as the emulsifier dosage increases. For WiDE with 9% water content, due to small amount of water, the emulsifier needs more surface area to attach on, hence SMD increases but at the same time more distribution of these bigger water particles is observed. In case of emulsions with 15 and 12% water content, the amount of emulsifier is enough to capsulate the water particles into smaller sizes as the emulsifier dosage increases. For the studied emulsions it can be concluded that when the water content is below 12%, the tendency to form larger water particle size is higher as the emulsifier dosage increases.

References

1. M.S. Kumar, J. Bellettre, M. Tazerout (eds.), *The Use of Biofuel Emulsions as Fuel for Diesel Engines: A Review* (SAGE Publications, Sage, 2009)
2. M. Abu-Zaid, Performance of single cylinder, direct injection diesel engine using water fuel emulsions. Energy Convers. Manag. **45**, 697–705 (2004)
3. O. Badrana, S. Emeishb, M. Abu-Zaidc, T. Abu-Rahmaa, M. Al-Hasana, M. Al-Ragheba, Impact of emulsified water/diesel mixture on engine performance and environment. Int. J. Therm. Environ. Eng. **3**, 1–7 (2011)
4. E.D. Scarpete, Diesel-water emulsion, an alternative fuel to reduce diesel engine emissions. A review. Mach. Technol. Mater. **7**, 13–16 (2013)
5. M. Yahaya Khan, Z. Abdul Karim, F.Y. Hagos, A.R.A. Aziz, I.M. Tan, Current trends in water-in-diesel emulsion as a fuel. Sci. World J. **2014**, 527–472 (2014)
6. M.Y. Khan, Z. Abdul Karim, A.R.A. Aziz, I.M. Tan, Experimental investigation of microexplosion occurrence in water in diesel emulsion droplets during the leidenfrost effect. Energy Fuels **28**, 7079–7084 (2014)
7. M. Yahaya Khan, Z. Abdul Karim, A.R.A. Aziz, I.M. Tan, Experimental study on influence of surfactant dosage on micro explosion occurrence in water in diesel emulsion. Applied Mechanics and Materials. **819**, 287–291 (2016)
8. M.E.A. Fahd, Y. Wenming, P. Lee, S. Chou, C.R. Yap, Experimental investigation of the performance and emission characteristics of direct injection diesel engine by water emulsion diesel under varying engine load condition. Appl. Energy **102**, 1042–1049 (2013)
9. V. Suresh, K. Amirthagadeswaran, Combustion and performance characteristics of water-in-diesel emulsion fuel. Energy Sources Part A-Recovery Util. Environ. Eff. **37**, 2020–2028 (2015)
10. I. Jeong, K.-H. Lee, J. Kim, Characteristics of auto-ignition and micro-explosion behavior of a single droplet of water-in-fuel. J. Mech. Sci. Technol. **22**, 148–156 (2008)
11. J. Ghojel, D. Honnery, K. Al-Khaleefi, Performance, emissions and heat release characteristics of direct injection diesel engine operating on diesel oil emulsion. Appl. Therm. Eng. **26**, 2132–2141 (2006)
12. M.Y. Khan, Z.A. Abdul Karim, A.R.A. Aziz, I.M. Tan, Performance and emission assessment of multi cylinder diesel engine using surfactant enhanced water in diesel emulsion. MATEC Web Conf. **13**, 02025 (2014)
13. M. Fingas, B. Fieldhouse, Formation of water-in-oil emulsions and application to oil spill modelling. J. Hazard. Mater. **107**, 37–50 (2004)
14. J. Jiao, D.J. Burgess, Rheology and stability of water-in-oil-in-water multiple emulsions containing Span 83 and Tween 80. AAPS J. **5**, 62–73 (2003)
15. Y. Zeng, F.L. Chia-fon, Modeling droplet breakup processes under micro-explosion conditions. Proc. Combust. Inst. **31**, 2185–2193 (2007)
16. M. Huo, S. Lin, H. Liu, F.L. Chia-fon, Study on the spray and combustion characteristics of water–emulsified diesel. Fuel **123**, 218–229 (2014)
17. L. Wang, J. Dong, J. Chen, J. Eastoe, X. Li, Design and optimization of a new self-nanoemulsifying drug delivery system. J. Colloid Interface Sci. **330**, 443–448 (2009)
18. G. Chen, D. Tao, An experimental study of stability of oil–water emulsion. Fuel Process. Technol. **86**, 499–508 (2005)
19. C.-Y. Lin, L.-W. Chen, Comparison of fuel properties and emission characteristics of two-and three-phase emulsions prepared by ultrasonically vibrating and mechanically homogenizing emulsification methods. Fuel **87**, 2154–2161 (2008)
20. C.-Y. Lin, K.-H. Wang, Diesel engine performance and emission characteristics using three-phase emulsions as fuel. Fuel **83**, 537–545 (2004)
21. H. Watanabe, T. Harada, Y. Matsushita, H. Aoki, T. Miura, The characteristics of puffing of the carbonated emulsified fuel. Int. J. Heat Mass Trans. **52**, 3676–3684 (2009)
22. V. Ganesan, *Internal combustion engines* (McGraw Hill Education (India) Pvt Ltd, India, (2012)
23. O. Armas, R. Ballesteros, F. Martos, J. Agudelo, Characterization of light duty diesel engine pollutant emissions using water-emulsified fuel. Fuel **84**, 1011–1018 (2005)

24. M.-G. Song, S.-H. Cho, J.-Y. Kim, J.-D. Kim, Novel evaluation method for the water-in-oil
    (W/O) emulsion stability by turbidity ratio measurements. Korean J. Chem. Eng. **19**, 425–430
    (2002)
25. G. Urbina-Villalba, M. García-Sucre, Brownian dynamics simulation of emulsion stability.
    Langmuir **16**, 7975–7985 (2000)

Chapter 5
Gasification of Date Palm (*Phoenix dactylifera*) Seeds

Mohammed Elamen Babiker, A. Rashid A. Aziz, Morgan Heikal and Suzana Yusup

5.1 Introduction

Biomass is the fourth biggest source of energy in the world, providing about 35% of the main energy in developing countries and around 3% in industrialized countries [1, 2]. In the last four decades, biomass energy became one of the most attractive subject of scientific research as it holds a great potentiality in replacing non-renewable fossil fuel [3, 4]. The kinetic study of the biomass pyrolysis is necessary to understand the behavior of energy during pyrolysis, gasification or combustion [5].

Many researchers have used the Aspen Plus simulator to measure mass and energy balances and to improve designs of biomass gasification process [6]. An atmospheric circulating fluidized bed (CFB) gasifier was used and simulated in a model developed by Doherty et al. [7]. Their study utilized Gibbs free energy minimization in combination with the restricted equilibrium method and validated it using experimental data. The method used specified temperatures of the gasifier reactions in order to predict

M. E. Babiker (✉) · A. R. A. Aziz
Department of Mechanical Engineering, Centre for Automotive Research and Electric Mobility, Universiti Teknologi Petronas, 32610 Seri Iskandar, Perak, Malaysia
e-mail: mohammed_g01834@utp.edu.my

A. R. A. Aziz
e-mail: rashid@utp.edu.my

M. Heikal
School of Computing, Engineering and Mathematics, Advanced Engineering Centre, University of Brighton, Brighton BN2 4GJ, UK
e-mail: m.r.heikal@brighton.ac.uk

S. Yusup
Department of Chemical Engineering, Universiti Teknologi Petronas, 32610 Seri Iskandar, Perak, Malaysia
e-mail: drsuzana_yusuf@utp.edu.my

© The Author(s), under exclusive licence to Springer Nature Singapore Pte Ltd., part of Springer Nature 2018
Z. A. Abdul Karim and S. A. Sulaiman (eds.), *Alternative Fuels for Compression Ignition Engines*, SpringerBriefs in Energy, https://doi.org/10.1007/978-981-10-7754-8_5

the composition of the product gas, heating value, and conversion efficiency. A wide range of temperature, air preheating and equivalence ratios were applied to achieve the objectives of their study. The results revealed that the product gas composition, conversion efficiency and heating value were significantly affected by the variation of the above-mentioned parameters. Moreover, their findings indicated the range of temperature and equivalence ratio (ER) in which high percentages of hydrogen and carbon monoxide are achievable. These results also showed that temperature and ER influenced the cold gas efficiency (CGE) and high heating value. Another study by Abdelouahed et al. [8] was to develop an original computer model of a preheating air CFB gasifier using Aspen Plus. However, their results showed that higher ER reduced the gas-heating value, while air preheating increased CO and H_2, and this in return increased the CGE and gas heating value. They reported the effectiveness of air preheating at lower ERs. They further revealed that the steam agent had a good reactivity compared to that of fuel-bound moisture, as high moisture degraded the gasifier operation. A pre-drying process of the feedstock is required to avoid a loss of system efficiency. They concluded that the presence of steam is necessary to obtain a syngas with a high percentage of H_2. A biomass gasification model using dual fluidized bed (DFB) reactors was studied by [8–11]. The researchers used Aspen Plus and dedicated Fortran files to develop the model. The bed was separated into three modules based on the main chemical occurrence, namely biomass pyrolysis module, secondary reactions module, and char combustion module. All [8–11] modelled permanent gases mass yields, species of 10 tar as well as char relating to the reactor temperature and pyrolysis correlation. Moreover, they modelled the second reaction using a semi-detailed kinetic approach to deal with the gas phase and catalytic conversion in the presence of CH_4 char and tar species, i.e. benzene, phenol, naphthalene and toluene, the water–gas shift reaction, soot–steam gasification and char. The findings revealed that experimental data obtained from the two bed technologies were in agreement with the calculated results of gas compositions, flow rates, and lower heating values. Additionally, they indicated that the water-gas shift reaction (WGSR) kinetic had a significant effect on the composition of gases and flow rates. They further optimized and reviewed WGSR and their kinetic laws. Another model was developed by Ramzan et al. [12] to investigate the behavior of the gasifier after reducing the moisture content and decomposing species into detailed yields. In their approach to predict the gasification process, they modelled the reactions using the Gibbs free energy minimization method. Temperature, biomass moisture content, equivalence ratio and steam injection were varied to study the effect of these parameters on the high heating value, cold gas efficiency and gas composition of biomass. They compared their simulation results with experimental data obtained from a hybrid biomass gasifier. The gasifier was fuelled with food, poultry and solid waste. The results showed that the increase in temperature increased the production of CO and H_2, while the increase in ER reduced the existence of CO and H_2 in the product gas and therefore reduced the CGE. The authors stated the moisture content as an influencing factor in terms of gas heating values. Moreover, they confirmed that the steam agent favors hydrogen production.

Table 5.1 Physical and chemical characteristics of date palm seeds

Samples	Weight percentage (%)	Bulk density (kg/m³)	Particle density (kg/m³)	Heating value (kJ/kg)
Deglet Nour	8.41	450	1451	18.548
Piarom	10.50	630	1524	18.238
Safawi	9.14	550	1511	18.226
Mabroom	8.86	530	1521	18.362
Aliya	12.56	730	1513	18.286
Suffry	8.33	500	1445	18.440

The current study was conducted to determine how date palm seeds (DPSs) feedstock would react to combustion process in a fluidized bed gasifier in order to conduct further investigations. Finally, one of the main purpose of this study is to develop a comprehensive simulation process of DPSs gasification in a pilot-scale fluidized bed gasifier to predict the syngas composition, performance and efficiency under different parameters and operating conditions. The study investigated the hydrodynamic and reaction rate kinetics, since thermodynamic equilibrium techniques are capable of measuring the effect of fuel and operating parameters on biomass. However, because the gasifier may not run under chemical or thermodynamic equilibrium, the maximum yield of desirable product gas using the Aspen Plus simulator without an optimized process model will give a reasonable prediction of gas compositions only. Therefore, a Fortran model kinetic created in Aspen Plus software is a useful tool to develop a model to predict the effect of hydrodynamic and geometric parameters such as design parameters and the fluidizing velocity of the gasification process.

5.2 Date Palm Seeds

Six types of DPSs cultivars were used in this analysis, namely Deglet Nour (DN) from Tunisia, Piarom (PI) from Iran, Safawi (SA), Mabroom (MA) and Suffry (SU) from Saudi Arabia, and Aliya (AL) from Algeria. They were obtained from a local importer in Penang city, Malaysia. All samples were thoroughly cleaned with distilled water and exposed to direct sunlight. Seeds dehydration process was carried out in two successive days with the aim of reducing total moisture content to an absolute minimum. Then, the samples were prepared by crushing, grinding and sieving the date seeds using a plier, ordinary home-type blender and ASTM Round Test Sieves, respectively. Sieves mesh sizes were No. 60, 250 μm and No. 45, 355 μm. Table 5.1 shows the types of date seed cultivars selected along with their physical and chemical properties.

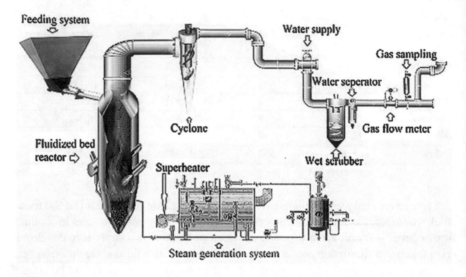

Fig. 5.1 Fluidized bed components for DPS steam gasification system

5.3 Experimental Set-Up and Procedures

Experimental works were conducted using date palm seed (DPS) as a feedstock in a pilot-scale gasifier for which the process flow diagram is shown in Fig. 5.1. The works were performed using a pilot-scale gasification plant for creating the numerical work to predict the behavior of the gasifier with different parameters and processes. Optimization of the parameters was taken into account to compute the overall and individual efficiencies of the gasifier's main components. The gasifier has two reactors (250 cm height, 15–20 cm internal diameter) made of Inconel alloys and contains a bed material and uses steam as a fluidizing agent. The main components of the DPS gasification process are displayed in Fig. 5.2. Six varieties of DPS from different regions, obtained from an importer in Penang, Malaysia, were characterized and tested in the gasifier. The DPSs were characterized for their ultimate and proximate analysis and the results were reported elsewhere [13].

5.4 Gasification Modelling

In this study, date palm seed samples from different regions were gasified. An Aspen Plus model was developed to predict the behavior of DPSs in a fluidized bed gasifier. Steady state operation under atmospheric conditions was assumed to build the simulation database. The kinetics reactions and hydrodynamics of the bed and freeboard were considered for an accurate prediction. The models were developed using experimental results obtained from the gasifier. As the process of gasification engaged

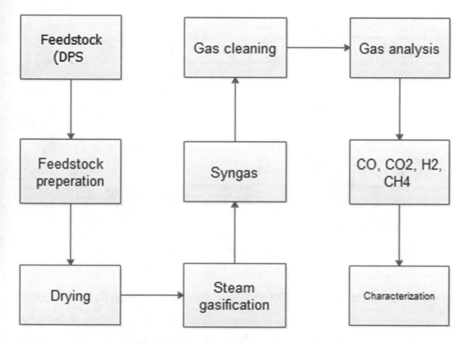

Fig. 5.2 Flow chart of DPS gasification process

many parts and zones to consider, the model developed for this particular study mainly focused on the process of gasification. Figure 5.3 shows the Aspen Plus simulation diagram of the bed gasifier. The bubbling fluidized bed is composed of two zones, i.e. a dense zone and freeboard zone, each with their own hydrodynamic properties. The dense zone is the place where the drying and devolatilization of the feedstock takes place and where superheated steam is injected at its lower boundary. Equations and correlation were used to calculate the local thermodynamic and hydrodynamics of the zone cells. However, the conservation equations of the carbon, bed material, and the energy of the whole of the zones were considered, instead of calculating the individual cells.

5.4.1   Modeling Primary Assumptions

In order to develop a model for predicting a steam-fluidized bed gasifier, the influence of hydrodynamic parameters on the DPS gasification in a fluidized bed and their kinetic reactions should be treated all together. The dense zone which represents the bubbling bed was modelled based on the modified two-phase theory, in which the bubbles size was considered as a function of the bed height. Moreover, all the bubbles were assumed to be of a uniform size. Therefore, the following valid assumptions

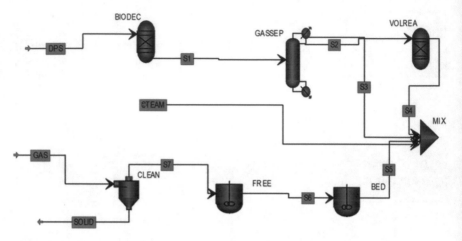

Fig. 5.3 Aspen Plus simulation diagram of fluidized bed gasifier

should be applied to the Aspen Plus simulator in order to simulate the real operation [14]:

1. According to the shrinking core model in which the particle size and the reacting core shrink simultaneously, the average particle size is constant and of a uniform size and spherical shape.
2. The distribution of gases within the emulsion phase (suspension of gas and solid around the bubbles and in the gasifier bed) is uniform.
3. Gasification is in the steady state, i.e. all state variables are constant regardless of ongoing processes.
4. The process is isothermal (uniform bed temperature).
5. DPSs have instantaneous de-volatilization compared to char gasification.
6. The product gas is composed mainly of CO, CO_2, H_2, CH_4 and water.
7. The volatile products mainly consist of CO, H_2, CO_2, CH_4 and H_2O.
8. DPS char starts in the bed, is completed in the freeboard, and contains C and ash compound.

5.4.2 Reaction Kinetics

The heat produced in the combustion process during the gasification supports a series of endothermic reactions. Therefore, combustible gases such as hydrogen, methane and carbon monoxide are obtained through these reactions in three consecutive processes, namely pyrolysis, combustion and steam gasification. The reactions that occur during these processes are [15].

$$C + \alpha O_2 \rightarrow 2(1 - \alpha) CO + (2\alpha - 1) CO_2 \qquad (5.1)$$

$$C + H_2O \rightarrow CO + H_2 \tag{5.2}$$

$$CO + H_2O \rightarrow CO_2 + H_2 \tag{5.3}$$

$$C + 2H_2O \rightarrow CO_2 + 2H_2 \tag{5.4}$$

$$C + \beta H_2O \rightarrow (\beta - 1)\,CO_2 + (2 - \beta)\,CO + \beta H_2 \tag{5.5}$$

where β is a mechanism factor as described by Nikoo and Mahinpey [16] with value of 0.5–1 for CO and CO_2, respectively, assuming that char combustion forces CO or CO_2 to leave the char particles.

As the char combustion is very slow process compared to that of volatilization, there is a sufficient time for the particles to spread around and burn in the bubbling bed. Char oxidation reaction takes place in the presence of oxygen and carbon on char surface to form CO and CO_2 [17]. Therefore, α is the link function between gasification temperature and diameter (average) of the char particles. The amount of steam that consumed during the reaction in Eq. (5.5) is calculated by $(2 - \beta)/\beta$, while the amount of steam that consumed during the reaction is found by $2(\beta - 1)/\beta$. Matsui et al. [18] managed to calculate β value as 1.5–1.1 when the temperature increase from 750 to 900 °C. In this study, and based on these values, β was found to be in the range of 1.4–0.9, which gave results in line with those obtained in the experimental work.

As the conversion process of gas and solid is assumed to be in the steady state, it is unreasonable to use analytical analysis for nonlinear rate equations in a bubbling-bed model [19]. Therefore, the numerical methods were developed in consecutive procedures to calculate the conversion. The methods used considered the level of solid conversion after its transit change, along with their corresponding gas phase concentration.

The mass transfer rate of particles is inversely proportional to their sizes according to the basics of mass transfer; on the other hand, the particle size has no effect on the reaction rate. Moreover, the size of char decreases with the advancement of combustion. Therefore, the kinetic rate remains independent, while a clear increase in mass transfer rate occurs. The following sequence of equations was described by Lee et al. [15], and modified by Nikoo and Mahinpey [16] to exemplify the reaction rate of biomass:

$$\frac{dX_{co}}{dt} = k_{co}\exp\left(\frac{-E_{co}}{RT}\right) P_{O_2}^n \, (1 - X_{co})^{2/3} \tag{5.6}$$

$$\frac{dX_{SG}}{dt} = k_{SG}\exp\left(\frac{-E_{SG}}{RT}\right) P_{H_2O}^n \, (1 - X_{SG})^{2/3} \tag{5.7}$$

$$rc = \left(\frac{dX_{CO}}{dt} + \frac{dX_{SG}}{dt}\right) \times \frac{\rho_C \varepsilon_S Y_C}{M_C} \tag{5.8}$$

where n equals 0.1 according to calculations made by Dutta and Wen [20]. However, some researchers reported different values for n; the actual value of n should be within the range of 0.9–1.0 in the steam partial pressure environment of 0.25 atm up to 0.8 atm [20].

5.4.3 Hydrodynamic Assumptions

To simulate the hydrodynamics in Aspen Plus reactors, the following assumptions were considered:

1. Two parts of the bed reactor, namely bed and freeboard must be simulated separately.
2. All fluidization regimes are running in bubbling condition.
3. According to the basic design of fluidized bed gasifier, solid particles decrease as the height increases. This phenomenon is explained by the bubbles coalescence at the bed zone and the return of particles at a transfer-disengaging height of the bed.
4. The height also positively affects the volumetric flow rate of gas in terms of yield.
5. The bed material, particles, char and ash are mixed flawlessly.
6. Hydrodynamic parameters of the reactor with a finite number of equal elements are considered.
7. No variation takes place in reaction conditions of gas and solid except for axial direction as the bed is a one-dimensional path.

5.4.4 Bed Hydrodynamics

The minimum fluidization velocity (umf) is the gas superficial velocity when the bed start fluidizing (the velocity at incipient fluidization). Ergun equation represents umf by calculating a pressure drop in a packed bed as the gas flow through it [21].

$$\frac{\Delta P}{h} = 150\frac{\left(1 - \epsilon_{mf}\right)^2}{\epsilon_{mf}^3}\frac{\mu_g u_{mf}}{\left(\emptyset_s d_p\right)^2} + 1.75\frac{1 - \epsilon_{mf}}{\epsilon_{mf}^3}\frac{\rho_g u_{mf}^2}{\emptyset_s d_p} \tag{5.9}$$

where ΔP is the pressure drop through the bed, with a positive value and h represents the length of a fixed bed. ρg is the density of the gas, while dp is the particle diameter. μg is the viscosity of the fluid. ϕs is the particle sphericity, where ϕs is the surface area of a sphere (having the same volume as a particle) divided by particle surface area.

When applying Ergun equation, the minimum fluidization voidage (ϵmf) is required to solve the problem. Therefore, ϵmf is usually considered within the range of 0.4–0.5. Researchers developed expressions for different particle types and sizes to be defined according to data obtained from experimental works. The descriptions are as follow:

$$\frac{1}{\emptyset_s\,\epsilon_{mf}^3} \cong 14 \quad \text{and} \quad \frac{1-\epsilon_{mf}}{\emptyset_s^2\,\epsilon_{mf}^3} \cong 11 \tag{5.10}$$

These descriptions modified for Ergun equation to obtain Reynolds number at ϵmf.

$$Re_{p,mf} = \frac{d_p u_{mf}\rho_g}{\mu_g} = \sqrt{C_1^2 + C_2 Ar} - C_1 \tag{5.11}$$

$$\text{where,} \quad Ar = \frac{d_p^3 \rho_g\,(\rho_s - \rho_g)\,g}{\mu_g^2} \tag{5.12}$$

where g represents standard gravity and ps is the density of the bed material. $C_1 = 33.7$ and $C_2 = 0.0408$, or instead $C_1 = 27.2$ and $C_2 = 0.0408$.

The following equations were introduced to calculate the minimum fluidization velocity for fine particles

$$u_{mf} = \frac{33.7\mu}{\rho_g d_p}\left(\sqrt{1 + 3.59 \times 10^{-5} Ar} - 1\right) \tag{5.13}$$

$$Ar = \frac{d_p^3 \rho_g\,(\rho_s - \rho_g)\,g}{\mu^2} \tag{5.14}$$

$$B = 1.0 + \frac{10.978\,(u - u_{mf})^{0.738}\,\rho_s^{0.376}\,d_p^{1.006}}{u_{mf}^{0.937}\,\rho_g^{0.126}} \tag{5.15}$$

$$\varepsilon_b = 1 - 1/B \tag{5.16}$$

$$\varepsilon_f = \varepsilon_b + (1 - \varepsilon_b)\,\varepsilon_{mf} \quad \varepsilon_{mf} = 0.4 \tag{5.17}$$

$$1 - \varepsilon_{fb} = (1 - \varepsilon_f)\exp(-az) \tag{5.18}$$

$$a = \frac{1.8}{u}. \tag{5.19}$$

5.5 Findings

Product gas composition of experimental and simulation data versus the three different temperatures (540, 730 and 800 °C) is depicted in Fig. 5.4. The results of H_2 revealed that the simulation findings are considerably in line with those obtained from experimental data. However, the increase in H_2 percentage at experimental (EXP) axis versus temperature (TEMP) is clear as it shows a sudden peak at the highest temperature (800 °C). Generally, DPSs were expected to show low percentages of H_2 at a lower temperature (540 °C), however, the predicted results revealed a tendency of this gas to form in nearly similar level compared to higher temperatures. Despite the fact that DPSs are highly reactive and described by high volatile content, this phenomenon could be a direct output of ignoring tar and unburned hydrocarbon

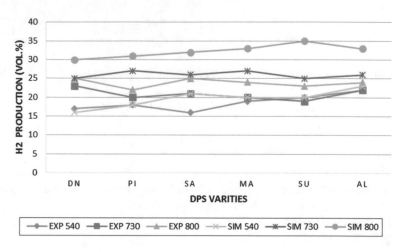

Fig. 5.4 Comparison between experimented and simulated effect of temperature on H_2 production (vol%)

formation at low temperature. This comparison is further demonstrated by the slope of the trend line to reflect the statistical trend of H_2 percentage in the two methods.

Figure 5.5 shows the predicted trend of CH_4 versus temperatures. It is clear that the simulation results correspond to that of experimental data. CH_4 decreased with the increase of temperature in almost all the varieties except for 730 °C where some disagreement to the general trend was observed at experimental curves. This decrease in CH_4 production could be due to the shift in the reaction of the methane reforming as higher temperatures act as an endothermic reformer, according to dynamic equilibrium, to produce a higher percentage of char and therefore lower CH_4 production [22].

The results shown in Fig. 5.6 present a related behavior to the production of carbon monoxide compared to the change in temperature. However, 730 and 800 °C temperatures were different from the experimental readings by predicting a regular pattern through all the curves of varieties except for DN and AL.

These different findings might be attributed to the uncontrolled treatments and parameters at high temperatures as well as control limits of gasifier at maximum running conditions. Moreover, DN and AL, are, to some extent, have different thermochemical characteristics in term of heating value and density compared to the other varieties.

The influence of temperature on carbon dioxide production is displayed in Fig. 5.7. The figure demonstrates the coherence between the simulated curves and the experimental data. Obviously, the predicted behavior of this compound showed that the formation of CO_2 favored the higher temperatures, which could be due to gasification backward reactions as the case with the experimental results [23]. This result might also be a spontaneous reaction to the oxidation of DPSs by oxygen elements associated with the steam agent. During the oxidation, a great amount of heat is released;

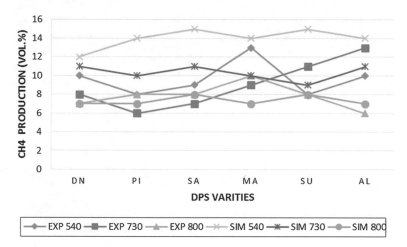

Fig. 5.5 Comparison between experimented and simulated effect of temperature on CH$_4$ production (vol%)

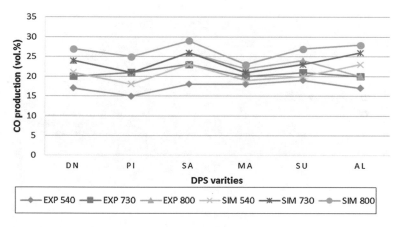

Fig. 5.6 Comparison between experimented and simulated effect of temperature on CO production (vol%)

however, CO will be produced as a result of partial oxidation that takes place due to a sub stoichiometric condition which generates CO. Furthermore, fuel conversion at lower temperatures is noticeable compared to higher temperatures (730 and 800 °C).

The results of carbon conversion efficiency and syngas production versus temperatures are displayed in Figs. 5.8 and 5.9, respectively. Fluidized bed gasifier has the lead of being very well mixed and having a high rate of heat transfer, resulting in uniform bed and freeboard conditions, and significant tar cracking [24]. However, higher temperatures are influencing factor too as the predicted and experimental charts confirmed the significance of this parameter on all varieties.

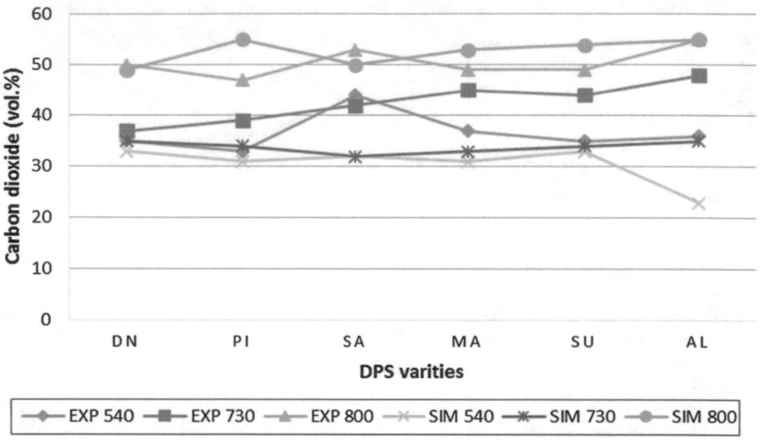

Fig. 5.7 Comparison between experimented and simulated effect of temperature on carbon dioxide production (vol%)

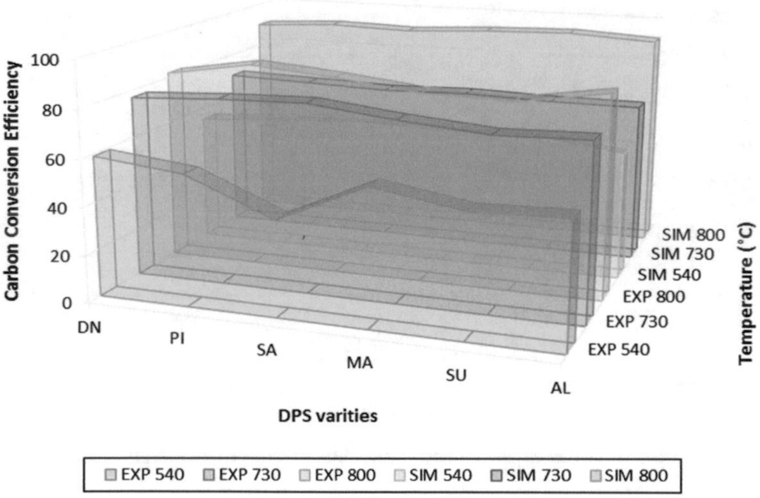

Fig. 5.8 Effect of temperature on carbon conversion efficiency (%)

In fact, unlike partial oxidation in which the gasification efficiency can reach up to 95–99%, the efficiency of the indirect-heat steam gasifier is only limited to 60–75% [25] in temperatures range from 790 to 870 °C. In this study, the results obtained for the carbon conversion efficiency, for both experimental and simulation methods, showed a good agreement with this fact as the range of temperature was intentionally extended to accommodate a lower temperature (540 °C), which is out of the ideal temperatures for bubbling fluidized bed gasifier. However, the lower temperature

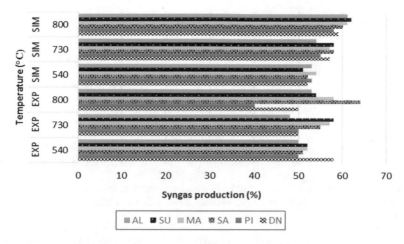

Fig. 5.9 Effect of temperature (°C) on syngas production (vol%)

favors the formation of CO, which is an advantage to be considered concerning syngas quality.

From all previous results, it was evident that using date palm seeds in a fluidized bed gasifier with steam as a gasifying agent at higher temperature (above 700 °C) and 1-1 S/B mass ratio can significantly increase the efficiency of the overall process. The pyrolysis temperature, namely 540 °C, was found to have a significant influence on determining the amount of combustible gases. The highest yield of H_2 (35%) was obtained for the SU variety at a simulated 800 °C, while the lowest yield (16%) was observed for both DN and SA, at experimental and simulated 540 °C temperatures. Methane production reached the peak (15%) under the simulated 540 °C and dropped to the lowest level (6%) and (7%) when the temperature was set to an experimental and simulated value of 800 °C, respectively. Moreover, variation of CH_4 production with temperature was evident for all varieties in both the experimental and simulated methods. The maximum production (29%) of CO was recorded at a simulation temperature of 800 °C; however, the percentage difference with the experimental level (26%) represents only a 10% reduction. The lowest value difference between CO productions (15%) was nearly 16% lower for the simulated 540 °C temperature compared to 18% of the experimental 540 °C temperature. Maximum CO_2 was recorded at 55% production, for the PI and AL varieties under the simulated 800 °C temperature, while the lowest value for CO_2, 23%, was observed at the simulated 540 °C only. In summary, the gas yield, lower heating value, carbon conversion efficiency and steam decomposition improve considerably with the increase of temperature. Finally, the current study introduced a new type of biomass waste, DPS, as a promising source of syngas production. The methods employed can be further adapted to predict the behavior of a wide range of gasifier feedstocks.

References

1. A.M. Omer, Built environment: relating the benefits of renewable energy technologies. Int. J. Auto. Mech. Eng. (IJAME) **5**, 561–575 (2012)
2. M.F. Demirbas, M. Balat, H. Balat, Potential contribution of biomass to the sustainable energy development. Energy Convers. Manage. **50**(7), 1746–1760 (2009)
3. J.S. Lim, Z.A. Manan, S.R.W. Alwi, H. Hashim, A review on utilisation of biomass from rice industry as a source of renewable energy. Renew. Sustain. Energy Rev. **16**(5), 3084–3094 (2012)
4. M. Mohammed, A. Salmiaton, W.W. Azlina, M.M. Amran, A. Fakhru'l-Razi, Y. Taufiq-Yap, Hydrogen rich gas from oil palm biomass as a potential source of renewable energy in Malaysia. Renew. Sustain. Energy Rev. **15**(2), 1258–1270 (2011)
5. M. Balat, M. Balat, E. Kırtay, H. Balat, Main routes for the thermo-conversion of biomass into fuels and chemicals. Part 1: pyrolysis systems. Energy Convers. Manage. **50**(12), 3147–3157 (2009)
6. A.J. Keche, A.P.R. Gaddale, R.G. Tated, Simulation of biomass gasification in downdraft gasifier for different biomass fuels using ASPEN PLUS. Clean Technol. Environ. Policy **17**(2), 465–473 (2015)
7. W. Doherty, A. Reynolds, D. Kennedy, Simulation of a circulating fluidised bed biomass gasifier using ASPEN Plus: a performance analysis, conference papers (2008)
8. L. Abdelouahed, O. Authier, G. Mauviel, J.-P. Corriou, G. Verdier, A. Dufour, Detailed modeling of biomass gasification in dual fluidized bed reactors under Aspen Plus. Energy Fuels **26**(6), 3840–3855 (2012)
9. H. Al-Kayiem, Y. Md Yunus, Drying of empty fruit bunches as wasted biomass by hybrid solar–thermal drying technique. J. Mech. Eng. Sci. **5**, 652–61 (2013)
10. N. Muda, M. Boosroh, Gasification of coal-petcoke blends in a pilot scale gasification plant. Int. J. Auto. Mech. Eng. **8**, 1457 (2013)
11. S.A. Sulaiman, N. Mat Razali, R.E. Konda, S.M. Atnaw, M.N.Z. Moni, On the diversification of feedstock in gasification of oil palm fronds. J. Mech Eng. Sci. **6**, 907–915 (2014)
12. N. Ramzan, A. Ashraf, S. Naveed, A. Malik, Simulation of hybrid biomass gasification using Aspen plus: a comparative performance analysis for food, municipal solid and poultry waste. Biomass Bioenerg. **35**(9), 3962–3969 (2011)
13. M.E. Babiker, A.R.A. Aziz, M. Heikal, S. Yusup, Pyrolysis characteristics of Phoenix dactylifera date palm seeds using thermo-gravimetric analysis (TGA). Int. J. Environ. Sci. Dev. **4**(5), 521 (2013)
14. S.S. Sadaka, A. Ghaly, M. Sabbah, Two phase biomass air-steam gasification model for fluidized bed reactors: Part I—model development. Biomass Bioenerg. **22**(6), 439–462 (2002)
15. J.M. Lee, Y.J. Kim, W.J. Lee, S.D. Kim, Coal-gasification kinetics derived from pyrolysis in a fluidized-bed reactor. Energy **23**(6), 475–488 (1998)
16. M.B. Nikoo, N. Mahinpey, Simulation of biomass gasification in fluidized bed reactor using ASPEN PLUS. Biomass Bioenerg. **32**(12), 1245–1254 (2008)
17. K. Jayaraman, I. Gokalp, Effect of char generation method on steam, CO_2 and blended mixture gasification of high ash Turkish coals. Fuel **153**, 320–327 (2015)
18. I. Matsui, D. Kunii, T. Furusawa, Study of fluidized bed steam gasification of char by thermo-gravimetrically obtained kinetics. J. Chem. Eng. Jpn. **18**(2), 105–113 (1985)
19. K.J. Timmer, *Carbon Conversion During Bubbling Fluidized Bed Gasification of Biomass* (Iowa State University, Ames, 2008)
20. S. Dutta, C. Wen, Reactivity of coal and char. 2. In oxygen-nitrogen atmosphere. Ind. Eng. Chem. Process Des. Dev. **16**(1), 31–37 (1977)
21. K. Kunii, O. Levenspiel, *Fluidization Engineering*, 2nd edn. Elsevier (2013)
22. K. Liu, C. Song, V. Subramani, *Hydrogen and Syngas Production and Purification Technologies* (Wiley, London, 2009)
23. S. McAllister, J.-Y. Chen, A.C. Fernandez-Pello, *Fundamentals of Combustion Processes* (Springer, Berlin, 2011)

24. G. Liu, P. Vimalchand, W.W. Peng, *Second Stage Gasifier in Staged Gasification and Integrated Process* (Southern Company Services, Inc., Birmingham, AL, (United States), 2015)
25. C. Stevens, R.C. Brown, *Thermochemical Processing of Biomass: Conversion into Fuels, Chemicals and Power* (Wiley, London, 2011)

Printed in the United States
By Bookmasters